Engineers, Society, and Sustainability

Synthesis Lectures on Engineers, Technology, and Society

Editor
Caroline Baillie, *University of Western Australia*

The mission of this lecture series is to foster an understanding for engineers and scientists on the inclusive nature of their profession. The creation and proliferation of technologies needs to be inclusive as it has effects on all of humankind, regardless of national boundaries, socio-economic status, gender, race and ethnicity, or creed. The lectures will combine expertise in sociology, political economics, philosophy of science, history, engineering, engineering education, participatory research, development studies, sustainability, psychotherapy, policy studies, and epistemology. The lectures will be relevant to all engineers practicing in all parts of the world. Although written for practicing engineers and human resource trainers, it is expected that engineering, science and social science faculty in universities will find these publications an invaluable resource for students in the classroom and for further research. The goal of the series is to provide a platform for the publication of important and sometimes controversial lectures which will encourage discussion, reflection and further understanding.

The series editor will invite authors and encourage experts to recommend authors to write on a wide array of topics, focusing on the cause and effect relationships between engineers and technology, technologies and society and of society on technology and engineers. Topics will include, but are not limited to the following general areas; History of Engineering, Politics and the Engineer, Economics , Social Issues and Ethics, Women in Engineering, Creativity and Innovation, Knowledge Networks, Styles of Organization, Environmental Issues, Appropriate Technology

Engineers, Society, and Sustainability
Sarah Bell
2011

A Hybrid Imagination: Science and Technology in Cultural Perspective
Andrew Jamison, Steen Hyldgaard Christensen, and Lars Botin
2011

Engineers, Society, and Sustainability

Sarah Bell

ISBN: 978-3-031-00982-2   paperback
ISBN: 978-3-031-02110-7   ebook

DOI 10.1007/978-3-031-02110-7

A Publication in the Springer Nature series
SYNTHESIS LECTURES ON ADVANCES IN AUTOMOTIVE TECHNOLOGY

Lecture #17
Series Editor: Caroline Baillie, *University of Western Australia*
Series ISSN
Synthesis Lectures on Engineers, Technology, and Society
Print 1933-3633   Electronic 1933-3633

Engineers, Society, and Sustainability

Sarah Bell

University College London and University of Western Australia

SYNTHESIS LECTURES ON ENGINEERS, TECHNOLOGY, AND SOCIETY #17

ABSTRACT

Sustainable development is one of the key challenges of the twenty-first century. The engineering profession is central to achieving sustainable development. To date, engineering contributions to sustainability have focused on reducing the environmental impacts of development and improving the efficiency of resource use. This approach is consistent with dominant policy responses to environmental problems, which have been characterised as ecological modernisation. Ecological modernisation assumes that sustainability can be addressed by reforming modern society and developing environmental technologies. Environmental philosophers have questioned these assumptions and call into question the very nature of modern society as underlying the destruction of nature and the persistence of social inequality. Central to the crises of ecology and human development are patterns of domination and the separation of nature and culture.

Engineering has a clear role to play in ecological modernisation, but its role in more radical visions of sustainability is uncertain. Actor-network theory provides an analysis of socio-technical systems which does not require the separation of nature and culture, and it provides a way of thinking about how engineers are involved in shaping society and its relationship to the environment. It describes the world in terms of relationships between human and non-human actors. It shows that social relationships are mediated by technologies and non-human nature, and that assumptions about society and behaviour are 'baked-in' to technological systems.

Modern infrastructure systems are particularly important in shaping society and have significant environmental impacts. Modern infrastructure has allowed the consumption of resources far beyond basic human needs in developed countries. Failure to deliver infrastructure services has resulted in billions of the world's poorest people missing out on the benefits of modern development. Engineers have an important role to play in developing new infrastructure systems which acknowledge the relationships between technology and society in shaping demand for resources and environmental impacts, as well as alleviating poverty.

Engineers have an important role in mediating between the values of society, clients, the environment and the possibilities of technology. Constructive Technology Assessment and Value Sensitive Design are two methodologies which engineers are using to better account for the social and ethical implications of their work. Understanding engineering as a hybrid, socio-technical profession can help develop new ways of working that acknowledge the importance of technology and infrastructure in shaping social relationships that are central to achieving sustainability.

KEYWORDS

sustainability, actor-network theory, consumption, ecological modernisation, infrastructure, water, socio-technical systems, environmental ethics

Contents

Acknowledgments

This book is based on the advanced level module 'Systems, society and sustainability' which I taught in the Department of Civil, Environmental and Geomatic Engineering at University College London 2006-2010. The ideas have been developed and refined in conversation with the students, teaching assistants and guest lecturers who contributed to the course over that time.

The book was written during my time as a Visiting Researcher in the School of Environmental System Engineering at the University of Western Australia, who provided much needed practical and intellectual support.

Sarah Bell
August 2011

Introduction

In the original charter of the Institution of Civil Engineers in 1828, Robert Tredgold defined engineering as the 'art of directing the great sources of power in nature for the use and convenience of man.' Industrialisation and modernisation, based on the dominance of humans over nature, has improved the lives of millions of people, but at the expense of environmental degradation and with uneven distribution of social costs and benefits. Engineers have been central to the process of industrialisation and responsible for the delivery of the technologies and infrastructures that define modern life. Engineers are therefore implicated in the ecological destruction and social disruption that have been the unintended consequences of industrialisation. Engineers are also essential for solving the problems of sustainability, and many engineers now understand their brief to include directing the powers of humanity for the benefit of nature, the flipside of Tredgold's definition. The contradictory depictions of engineers controlling and destroying nature and communities, or protecting and restoring nature and achieving sustainability are not easily reconciled.

Engineering is a powerful profession, marshalling science, technology and practical know-how to drive the processes of economic development and material progress. The role of engineers in setting the direction of change is less clear. Engineers have an ambivalent relationship with politics and society. On the one hand, engineers are conceived as servants of political and commercial masters, implementing decisions taken beyond their control. On the other hand, engineers are powerful social and political agents responsible for modern revolutions such as the provision of clean drinking water, space exploration and personal computing.

Sustainable development and sustainability are contested concepts with multiple definitions, but they generally require consideration of social, economic and ecological factors together in decisions and action. The role of engineers in creating and solving the crises of sustainability is ambiguous, but in order to achieve sustainability, engineers need to work with nature and society to build technologies and systems that restore and maintain essential ecological processes and deliver the benefits of development to billions of people currently living in poverty. Engineers have made considerable contributions to improving the environmental performance of industries, technologies and infrastructure, but they have been less successful in positively influencing the social dimension of sustainability. They have largely focussed on improving energy and resource efficiency and reducing waste and pollution. The next step in sustainable engineering is to move beyond efficiency and reducing environmental impacts towards a clearer recognition of the role of engineers in shaping society as well as building technology.

In taking the next step towards sustainable engineering, it is helpful to have some knowledge of how we have reached this point and a map of the obstacles and path ahead. In this book, I will critically reflect on the nature of engineering and its role in modern society in order to better understand the contradictions between engineering as the cause of environmental problems and the source of sustainable solutions. I will explore theories about the relationships between technology and society to propose new models of engineering practice that allow more integrated, sustainable practice, and provide pointers for the direction of change in the profession and its position in society. The core question I seek to answer is the following: *How can engineers help build a sustainable society?*

ENGINEERING SUSTAINABILITY

In the past decade, considerable gains have been made in incorporating sustainability into the everyday expectations of professional engineering. At the highest levels of the profession definitions of sustainability acknowledge the need to address social as well as ecological issues. The 2006 'Sustainability Protocol' signed by the presidents of the Canadian Society for Civil Engineering, the American Society of Civil Engineers and the UK based Institution of Civil Engineers states that:

> ASCE, CSCE and ICE believe that the current approach to development is unsustainable. We are consuming the earth's natural resources beyond its ability to regenerate them. We are living beyond our means. This, along with security and stability, is the most critical issue facing our profession and the societies we serve.

> In addition to the environmental impacts of our actions, the needs of societies around the world are not being met. Our goal as civil engineers is the creation of sustainable communities in harmony with their natural environment. In doing so we will be addressing some of the most profound problems facing humanity, for example climate change and global poverty, to name only two (CSCE et al., 2006).

The most prominent attempts to bring sustainability and engineering together are based on an approach which incorporates ecological factors into conventional engineering models and techniques (Bell et al., 2011). Technical specifications for pollution control, for example, enable the environmental impact of engineered systems to be objectively monitored and maintained within predefined parameters. Resource efficiency can be characterised as an extension of conventional engineering concerns with efficiency in design, previously driven by cost concerns but now more explicitly by environmental factors (Hawken et al., 2000). Holistic approaches to engineering sustainability based on systems thinking connect scientific principles developed to describe ecological systems to engineering methodologies developed in manufacturing, chemical processing, computing and large project management. Biomimicry in design and industrial ecology provide the clearest examples of the direct translation of ecological knowledge and models into engineering practice (Allenby, 1999, Benyus, 2002, Yiatros et al., 2007).

Scientific and technical knowledge about ecological systems is more familiar to engineers than social and cultural issues and processes, and consideration of social and cultural dimensions of sustainability has been more limited in engineering practice. The Engineering Council UK (2009) in its 'Guidance on Sustainability' states that 'a purely environmental approach is insufficient, and increasingly engineers are required to take a wider perspective including goals such as poverty alleviation, social justice and local and global connections.' The social dimension plays a key role in conceptualising the suitability of engineered systems for different populations given cultural norms and economic and political realities. It also influences the success or failure of sustainability in more technical terms – for example, through the adoption or non-adoption of the sustainability agenda in particular fields or geographical regions, the need to match user expectations with system design, and predicting and shaping future consumption patterns.

ENGINEERING KNOWLEDGE

The difficulty of incorporating social considerations into engineering designs and decisions is in part due to the limitations of engineering knowledge in coming to terms with social realities. Ecological models and environmental data provide knowledge that is more-or-less compatible with engineering models and decision making. The environment can be known in quantitative terms that are familiar to engineers, and engineering has been relatively successful in developing tools for improving ecological efficiency and reducing environmental impact. Social life is not as easily abstracted into sets of numbers or equations that can connect with engineering systems models or design codes. Knowledge from the social sciences is often qualitative, and where social phenomena are quantified, the data are usually specific to a particular population and less likely to be universally transferrable than the physical phenomena that engineers conventionally work with.

Enhancing engineers' contribution to sustainability as a social as well as ecological challenge requires a reconsideration of the nature of engineering knowledge along with improved understanding of how society and technology interact. Despite significant gaps between engineering and social science world-views, there is scope within engineering knowledge and practice to better understand social phenomena and draw on this knowledge to devise more sustainable engineering systems.

Engineering knowledge has always been based on collective experience of what works in solving particular problems (Vermaas et al., 2011). Experimental methods and knowledge from the sciences have contributed to advancing engineering knowledge, but engineering is far from the mere application of scientific knowledge. Engineers' attention to the specifics of each particular site or design problem, the importance of 'rules of thumb,' the necessity to identify and choose between possible solutions rather than a searching for one absolute answer and the high value placed on creativity in devising innovative solutions are all compatible with social science knowledge and methods.

Engineering knowledge develops from solving particular problems. Sustainability is now widely recognised as one of the core problems which the profession must address, including social as well as environmental considerations. Engineering is increasingly looking to incorporate knowledge from the social sciences. This presents a challenge to orthodox quantitative engineering methods that are most familiar with the physical and biological sciences, but the engineer's capacity to draw on collective experience and specific evidence in solving complex problems provides a good foundation for expanding the engineering sciences to include knowledge of society and everyday life.

SOCIAL SUSTAINABILITY

Parallel to the developments in engineering sustainability, the social sciences and humanities have also made significant progress in helping to understand the environmental crisis and its relationship to society. Environmental sociology, psychology, anthropology, human geography, philosophy, politics and public policy, economics, development studies and planning are but a few dimensions of the vast contribution made by social scientists to better understand the challenge of sustainability. Social scientists use a wide range of methods and theories to understand environmental and sustainability problems from different and sometimes contradictory perspectives. Whilst much of this knowledge is of relevance and interest to engineers, the most useful insights from the social sciences are likely to come from studies and theories which specifically focus on the relationships between technology and society. Social science which relates to the consumption of energy and water, the use of transport systems, household management of waste and other environmental activities as part of wider social and cultural phenomena are of particular interest to engineers working to design and manage sustainable systems and technologies. Public engagement and participation in decision making are also vital to achieve sustainability, and an important area in which the social sciences can inform engineering.

The conventional division between society on one side, and science and technology on the other is central to understanding the problems of sustainability. Science and society, or nature and culture, have been thought of as completely separate domains, with modern technology often assumed to be merely applied science. Science and technology have been left to the engineers and scientists to develop at an objective distance from society and the messy worlds of politics and culture. Some scholars have traced the origins of the modern environmental crisis to this division between nature and culture, corresponding to science and society (Plumwood, 1993). Others have pointed out that the division has never been as clear cut as assumed, and that science, technology, society and politics have always been tied together in complex relationships (Latour, 1993).

When investigating environmental problems and working toward sustainability, it soon becomes apparent that dividing problems into social, scientific and technological components is highly fraught. Actor-network theory (ANT) developed from social and anthropological studies of science and technology and is useful in helping to understand social, natural and technological elements of any system on the same terms. ANT does not distinguish between society, technology or nature, but instead describes the world in terms of human and non-human actors which form networks of relationships amongst one another. A methodology which does not distinguish between social and technological phenomena is a useful standpoint for better understanding sustainability problems and designing solutions, and it provides a common platform for engineers and social scientists to engage with knowledge from different disciplines and world-views.

ENGINEERING, SOCIETY AND SUSTAINABILITY

In answering the question *how can engineers help build a sustainable society?* this book will analyse the problems of sustainability and the nature of engineering in terms of complex relationships between technology, society and the environment. Chapter 1 outlines the problems of the ecological crisis and the origins of the concept of sustainable development. Chapter 2 provides an overview of the most common tools engineers use to address sustainable development, and places them in the context of the dominant policy approach which has been characterised as ecological modernisation. Chapter 3 investigates philosophical critiques of modern society and culture in light of the ecological crisis, pointing to the need to fundamentally reconsider the relationships between humans and the natural world and the role of engineers. Chapter 4 presents the actor-network theory as a key contribution from social science that highlights the importance of relationships between humans and non-humans in society, including the role of technology and the non-human natural world. Chapter 5 considers the importance of technology and infrastructure in shaping unsustainable patterns of consumption of energy, water and materials. Chapter 6 analyses case studies of urban water systems, with a particular attention to the strengths and weaknesses of engineering approaches to solving this vital problem. Chapter 7 explores the role of values in technology and engineering, and presents value-sensitive design and constructive technology assessment as methods for making values and social assumptions more explicit in engineering. The book concludes by summarising the key challenges of sustainability and outlines a theory of engineering practice which is better able to respond to them.

CHAPTER 1

The Origins of Sustainability

Sustainable development aims to halt and reverse the degradation of global ecosystems whilst maintaining human development. Given that human development has conventionally been based on industrialisation and modernisation, which in turn have required the exploitation of natural resources and the pollution of the environment, sustainable development is seen by many as an oxy-moron, essentially a contradiction in terms. Alternatively, others understand sustainable development as the greening of industrialisation. They believe that sustainable development can be achieved by taking better account of the environment within the processes of industrialisation and development as we know it.

In order to engage in debate about the nature of sustainable development and the role of engineers, it is important to understand the basic problems that it seeks to address. The magnitude of these problems is daunting. The human impact on global ecosystems has been vast, and billions of people still do not have access to the basic benefits of development such as sanitation, health care or primary school education. This chapter addresses the nature of these problems and provides an outline of the history of how the ecological crisis and development have been addressed by the global community, leading to the emergence of sustainable development as a vital compromise between two seemingly conflicting movements.

1.1   GLOBAL ECOLOGICAL CRISES

The human impact on the environment is a multi-dimensional problem. It is a problem both for the environment itself and for the future of humanity. The growth in human population and consumption is undermining the capacity for the environment to provide essential services such as clean air and water, and a stable climate in which to live and undertake the complex tasks of modern life. In 1997, the economic value of such 'ecosystem services' was estimated to be US$33 trillion per annum (Costanza et al., 1997). We are consuming renewable resources such as water, timber and food at rates that are higher than the ability of the environment to replenish them, risking collapse of these essential life support systems. Consumption of non-renewable resources such as oil and minerals without adequate attention to future needs for these resources or development of alternatives risks the collapse of economic and social systems. Extraction, processing and use of resources from the environment through mining, harvesting, refining, manufacturing and waste disposal result in direct destruction and pollution of ecosystems, undermining their ability to function. Whilst the consequences of these impacts for people, now and in the future, are significant and important, the loss of natural places and the impact of human activities on other life forms are moral issues in themselves. It may be justifiable to exploit other species and modify the environment to meet basic human needs, but the widespread destruction of natural systems to support ever expanding rates of consumption and continually rising material standards of living is not as easily reconciled. This section provides a brief review of a selection of environmental issues. It is by no means comprehensive and does not address important issues such as the state of the world's fisheries, wildlife conservation, chemical pollution, and water quality.

1.1.1   POPULATION

The world population in 1800 was approximately 1 billion people. It is currently around 7 billion people and projected to stabilise at around 9 billion by 2050 (UNDESA, 2004). Population is growing fastest in least developed countries. In 2007, for the first time in history, more than 50% of the world's population lived in cities and more than 1 billion people lived in urban slums (UN Habitat, 2006).

Population growth rates are linked to human development through what is known as the 'demographic transition.' As industrialisation begins, death rates are reduced but birth rates remain high, leading to rapid growth in population size. As the benefits of industrialisation are stabilised and become more widespread, birth rates decline and population stabilises. If birth rates decline below the replacement rate, populations can start to shrink. Some key factors which help reduce birth rates are: good maternal and child health; compulsory education; improved education and economic status of women; and access to good family planning services. Improving child and maternal health increases the likelihood of babies surviving until adulthood, reducing the need for 'insurance births.' Compulsory education shifts children from being economic resources for poor families to being economic costs. Improved education and access to paid work increase the status of women within society and families, providing opportunities beyond the household and the capacity to have more control over reproduction. Access to safe, reliable and dignified family planning services further supports people's desire to limit the size of their families as economic and social conditions change.

The growing global population places increasing pressure on natural resources and the environment. Not only is the population growing, but per capita consumption levels are also increasing. For some people, higher levels of resource consumption are required to meet basic human needs. For others, higher levels of consumption are associated with an increasingly luxurious lifestyle.

1.1.2   WATER

Water is essential for life. Freshwater is a renewable but limited resource. 70% of the Earth's surface is covered in water, but less than 3% of the water on the planet is freshwater, and only 0.6% of all freshwater or less than 0.01% of total water is available for human use. Growing population and growing consumption place local water resources under severe stress in a number of areas of the world. Throughout the twentieth century, water use grew at twice the rate of population growth (UNCSD, 1999). This is primarily due to irrigation for agricultural production, but it also reflects growing industrial and domestic consumption. Agriculture accounts for approximately 70% of global water use, with industry consuming 19% and domestic users consuming 11% (Table 1.1). Demand for

My family's demographic transition

Fertility is often assumed to be a highly personal and culturally and religiously sensitive issue, making family sizes difficult to change. However, there is a strong link between women's status and education and fertility, which can be illustrated by changes in fertility across generations of my own family. My family is Catholic and fifth generation Australian.

My grandfather was one of 11 children, nine of whom survived to adulthood. His mother had a primary school education. My mother was one of seven children, and her mother had a high school education. I am one of four children and my mother had a two year post-high school teacher's certificate before having children. I have a PhD, and no children.

water is predicted to continue to grow to meet demand for food production as well as increased domestic consumption. By 2025, two thirds of the world's population are predicted to be living under conditions of water stress, where demand for water exceeds the capacity of the environment to meet it (UN Water, 2006).

Table 1.1: Selected International Water Use (FAO, 2010, Gleik, 2008)

	Agriculture (%)	Industry (%)	Domestic (%)
Global Average	70	19	11
Afghanistan	98	0	2
Australia	75	10	15
Singapore	4	51	45
UK	3	75	22
USA	41	46	13

Domestic consumption of water varies widely around the world (Figure 1.1). Improving access to safe drinking water is essential for good public health, and overconsumption of water places local resources and aquatic ecosystems under stress. In 2015, it is likely that one in ten, or 700 million, people will not have safe access to water within one kilometre of their home (UN, 2011). By contrast, in 2002, the average per capita domestic consumption of water in the USA was more than 500L, in Australia was more than 450L and in the United Kingdom was 150L per day (Figure 1.1). Increasing basic access to water in the developing world and managing profligate demand in the developed

world are both major challenges for engineers working to address sustainability, and they will be
further explored in Chapter 6.

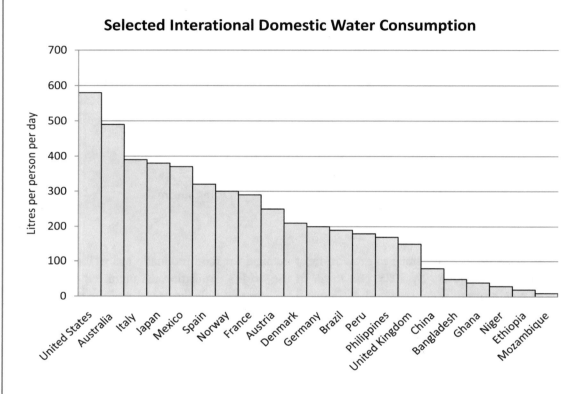

Figure 1.1: International domestic water use (United Nations Development Programme, 2006).

1.1.3   SOIL

The rapid rise in global population has only been possible because of dramatic increases in food
production since the nineteenth century, particularly in the second half of the twentieth century.
The 'green revolution' of the nineteen sixties and seventies, based on the expansion of irrigation, the
application of industrial fertilisers and pesticides, mechanisation and selective breeding species of
crops and livestock, resulted in dramatic increases in food production. However, the environmental
impacts of intensive agricultural production have been significant, most importantly in relation to soil
degradation. Soil erosion, salinisation, acidification, compaction and other processes of degradation
and desertification affect significant and growing areas of the world's arable land (UNEP, 2007).
Analysis of net primary productivity, or biomass production, between 1981 and 2003 showed an
absolute reduction across of 12% of the Earth's land surface. Of the degraded land, 18% is associated

with cropland and 42% is associated with degradation or loss of forests (UNEP, 2007). Soil degradation and desertification are linked to poverty, as poorer farmers are more likely to over-exploit limited land resources and are more vulnerable to the impacts of soil degradation. Nearly 250 million people are directly affected by desertification (Johnson et al., 2006). Protecting and restoring the world's soil resources is essential for the future of food production. Soil conservation techniques draw on both traditional and modern agricultural knowledge and include reducing over-irrigation, minimising tillage and compaction, and restoring trees and other perennial vegetation within the agricultural landscape.

1.1.4   DEFORESTATION

The development of land for agriculture has been one of the main causes of deforestation, along with logging. Between 1990 and 2005, the world's forest cover declined overall by 0.2% each year, with increases in North America, Europe, Asia and the Pacific making up for higher rates of forest loss in Africa, Latin America and the Caribbean (UNEP, 2007). The quality of forests is also changing, with higher losses of primary, or old growth, forests, and increases in plantation forest, which provide fewer ecosystem services. Forests are managed for a range of uses including timber production, fuel and ecosystem services. Loss of forests results in a decline in stocks of renewable resources such as timber, increases soil degradation and flood risk due to increased runoff, and has a significant impact on global biodiversity.

Green Belt Movement

The Green Belt Movement was started in Kenya in 1977 by Dr. Wangari Maathai to help women in rural areas plant trees for fuel and soil and water conservation. The organisation established tree nurseries and facilitated community development to enable women to lead conservation activities and sustainable development in their local area. Since then, more than 40 million trees have been planted and the movement has expanded across Africa and now includes a global programme. Dr. Maathai was awarded the 2004 Nobel Peace prize in recognition of these achievements (Green Belt Movement, 2006).

1.1.5   BIODIVERSITY

The human impact on the abundance and diversity of other species has been immense. The current rate of species extinction due to human intervention in the environment has been compared with previous extinction events caused by dramatic geological and meteorological changes in the Earth's history (Reaka-Kudla et al., 1997, Wilson, 1992). The International Union for the Conservation of Nature estimates that 1 in 3 amphibians, 1 in 4 coniferous trees and mammals, and 1 in 8 bird species are vulnerable, endangered or critically endangered. Of 40,168 species assessed in their 2006 'Red List,' 16,118 were threatened and 65 were only found in captivity or cultivation (IUCN, 2006). Species extinction and loss of biodiversity result from deforestation, land degradation, disease, invasive species, climate change, direct exploitation and habitat loss. As such, the loss of biodiversity is a good indicator of wider processes of environmental degradation and ecosystem breakdown. Loss of biodiversity also removes opportunities for future discoveries of resources from plants and animals that could be useful to solve future problems of health, energy and food production. Furthermore, the extinction of other species as a result of human activities raises significant moral questions about our relationship with the non-human natural world.

1.1.6   OIL

Economic growth and industrialisation have been closely linked with increasing energy consumption. The early industrial revolution relied on hydropower to drive textile mills, and the steam powered revolution of the nineteenth century relied on coal. The abundance of cheap oil throughout much of the twentieth century transformed modes of transportation, agricultural production, international trade and urban development. As well as its energy value, oil also allowed the development of plastics, pesticides and other synthetic chemicals that have become central to modern economies and lifestyles.

Throughout most of the twentieth century, rates of discovery of new oil reserves exceeded the rate of growth in consumption, allowing oil production to increase to meet growing demand. Oil is a non-renewable resource and as rates of discovery of new reserves decline and consumption continues to grow, oil production will peak at some point in the coming decades (IEA, 2010). New technologies for extracting more oil from declining reserves and from non-conventional sources such as tar sands will extend the global oil resources, but they are likely to be more expensive than conventional production and may lead to higher carbon emissions (IEA, 2010). These technologies will become economically viable as their efficiency improves and as oil prices increase due to scarcity. Increasing scarcity of conventional oil is likely to contribute to increased oil prices over coming decades, with significant economic and social impacts. Industrial agriculture and international trade in foodstuffs are highly dependent on oil and food prices are highly vulnerable to oil prices. Adapting economies and societies to higher oil prices will require changes in urban and transportation planning, as well significant changes in international trade and industrial production.

1.1.7   CLIMATE CHANGE

Burning oil, gas and coal is the main cause of global climate change, which intensifies most other environmental problems as well as having direct impacts on the environment and human health and livelihoods. Burning fossil fuels and deforestation have increased the concentration of carbon dioxide in the atmosphere from pre-industrial levels around 280 parts per million by volume (ppmv)

Peak oil

Oil is a non-renewable resource. The volume of oil on Earth is finite. In the 1950s, the geologist Dr. M. King Hubbert proposed that oil production would increase until it reached a peak, after which time, it would decline until reserves were depleted. Production volumes observed over time in individual oil wells and aggregated to national oil production generally followed a logistic shaped curve - increasing rapidly to a peak and then declining. The peak is thought to be at the mid-point of exploitation of the total oil available for extraction. Hubbert's model has been applied successfully to predict the production peaks of individual wells and nations and has been scaled up to consider when global production of conventional oil will peak. These predictions account for known reserves and estimates of future discoveries based on past trends. Estimates for global peak production of conventional oil vary from 2006 to the 2020s.

to around 387 ppmv in 2008. This can be compared to the variation of between 180 and 300 ppmv over the last 650,000 years (Maslin, 2008). Increasing concentrations of CO_2 in the atmosphere have been linked to rising global surface temperature and sea levels over the same period. Sea levels have risen by 4cm and average global temperatures have risen by 0.76 degrees Celsius over the last 150 years (IPCC, 2007).

Predictions of future climate change are based on computer models of global climatic systems using different scenarios of climate change emissions. Global average temperatures could rise from between 1.1 degrees and 6.4 degrees by 2100 depending on levels of emissions, resulting in sea levels rising between 18cm and 59cm (IPCC, 2007). Increasing global temperature and sea level rise will have dramatic impacts on weather patterns around the world. Extreme weather events, such as heat waves and hurricanes, are likely to become more frequent. Rainfall patterns are likely to become more erratic, with increased likelihood of both floods and drought in some parts of the world. In 2009, the medical journal *The Lancet* identified climate change as the biggest threat to public health in the 21st century, both in terms of direct impacts due to extreme events and changing patterns of disease, and indirect impacts due to reduced food and water security (Costello et al., 2009). The Lancet Report highlighted that the poorest people in the world, those who contribute least to causing climate change, will suffer most from its effects.

1.1.8   ECOLOGICAL FOOTPRINT

With population growth rates highest in developing countries, and per capita consumption of resources highest in developed countries, comparison of the relative impacts of different countries and lifestyles on global ecosystems can be difficult. The concept of an ecological footprint allows for such comparisons to be made, highlighting the ecological impact of average lifestyles in different countries, as well as allowing for an assessment of the overall capacity of global ecosystems to support growing populations and levels of affluence. Ecological footprint is expressed in terms of hectares of the Earth's surface required to deliver the ecosystem services and resources needed to support a particular lifestyle and level of consumption. The average person living in North America requires 8 ha to support his or her lifestyle, while the average person living in Africa requires only 1.4 ha. Using these estimates, the North American population of 341 million people has a greater total ecological impact than 964 million Africans (Global Footprint Network, 2010).

1.2   THE ENVIRONMENTAL MOVEMENT

Concern about the impact of modernisation on the environment can be traced to the early years of the industrial revolution, but the modern environmental movement, responding to many of the problems outlined in the previous section, has its origins in the nineteen sixties. The early environmental movement focused on overpopulation, pollution and the protection of wilderness areas.

1.2.1   SILENT SPRING

One of the foundational texts of the environmental movement is *The Silent Spring* by Carson (2002). Carson was an ecologist and science writer who had worked for the US Bureau of Fisheries before becoming a full-time writer, largely focussed on marine biology and ecology. Her concern about the widespread use of synthetic pesticides led her to publish her most famous book *The Silent Spring* in 1962. The title of the book referred to a spring without birdsong, an outcome of the indiscriminate use of pesticides. Whilst chemical companies, governments and the general population were enthralled with the benefits of new chemical pesticides that killed unwanted insects in homes, streets and farms, Carson was one of the first to detail their environmental and health impacts. Her work was not only instrumental in ensuring better testing and control of synthetic pesticides, but it catalysed a wider social movement that began to question the assumption that the outcome of industrial development were inevitably positive, and highlight the environmental consequences of new developments in science, engineering and industrialisation.

1.2.2   NEO-MALTHUSIANS

A more controversial, but not less important, foundation for the modern environmental movement was the work of the 'neo-Malthusians' including Paul Ehrlich and Garrett Hardin. The 'Neo-Malthusians' were following a line of thought first developed by Thomas Malthus, an eighteenth century economist. Malthus observed rapidly increasing population during the industrial revolution in England and concluded that, left unchecked, population growth would outstrip growth in agricultural production leading to famine and the downfall of civilised society. Malthus did not account for the industrialisation of agriculture and the impact of new technologies such as mineral fertilisers, which increased agricultural production at rates faster than he predicted, thus averting the catastrophic food shortages and collapse of society which he had feared. However, by the nineteen sixties, with the global population growth rate at its highest ever levels, scientists again became concerned that growth in population would outstrip the ability of natural and industrial systems to produce the necessary resources, leading to mass starvation and ecological and social collapse.

Paul Ehrlich is an entomologist and conservation biologist at Stanford University who is best known for his 1968 book *The Population Bomb*. Ehrlich argued that exponential population growth, particularly in developing countries such as India, would outstrip agricultural production leading to famine and the collapse of ecological systems. He called for drastic population control measures to reduce fertility and population growth, forcibly if required.

Garrett Hardin was another biologist who wrote widely about the threat of overpopulation. His most famous paper, 'The tragedy of the commons,' was published in the journal *Science* in 1968. Hardin (1995) used the analogy of the collapse of an unregulated common land resource to argue for 'mutual co-ersion' to restrict the 'freedom to breed.' According to Hardin, common lands failed due to over grazing because individuals were not penalised for over stocking. When everyone over stocked, the resource collapsed and everyone suffered the consequences. For this reason he argued that it is necessary to restrict individual freedom in the interest of protecting common resources, and that this was particularly the case with regard to controlling fertility. Hardin argued that individuals would not choose to limit their fertility, and government intervention was required to constrain birth rates.

The neo-Malthusian position has been heavily criticised from different angles, but their warnings about the dangers of population growth remain prescient. Religious groups criticised the promotion of birth control, particularly abortion, as a means of limiting fertility. Economists, such as Julian Simon, criticised their pessimism in relation to technological innovation, which has allowed for continued growth in production and improved resource efficiency. They were criticised for focussing attention solely on population growth, which was largely a problem in developing countries, without considering the ecological impacts of high levels of consumption by more affluent, though stable, populations. Whilst some of their predictions were overly gloomy and some of their proposed solutions distasteful, neo-Malthusian warnings have inspired programmes to reduce birth rates and stabilise population.

1.2.3   LIMITS TO GROWTH

The idea of limits to growth in population and resource consumption was further developed in the 1970s through mathematical modelling and economic theory. The *Limits to Growth* report to the Club of Rome in 1972 was based on computer modelling of resources and pollution under different scenarios of population and consumption growth (Meadows et al., 1972). The research was led by

China's one child police

In 1949, the population of China was just over half a billion people and by 1981, it was 1 billion. Concern about population growth and its impact on Chinas ability to meet its economic development targets led to increasing attention on population control through the 1970s. Policies implemented in the early 1970s to delay marriage, provide free access to contraception, abortion and sterilisation and encourage longer spacing between births, together with economic and social change, resulted in a reduction of urban birth rates from 3.3 births per woman in 1970 to 1.5 in 1978. Over the same period rural birth rates declined from approximately 6 to 3 births per woman. Despite the remarkable success of these policies in rapidly reducing birth rates, the overall birth rate was still considered too high, and so in 1979, China's leaders implemented the drastic policy of only allowing one child per couple (White, 2003). The implementation of the policy involved strict surveillance of women and strong penalties for unauthorised births, including loss of employment and fines. Regional family planning authorities were set targets for numbers of births as well as numbers of sterilisations and abortions, which they met through strategies of encouragement, coercion and surveillance. The policy has met with resistance, particularly in rural areas. Over time, the policy in some areas shifted to allow couples a second pregnancy if the first child was a girl, so they could try again for a boy, and further relaxed to allow two children of any sex for some instances. A social preference for boys resulted in sex selection during pregnancy and some cases of infanticide of girls, resulting in 117 boys born for every girl in 1995, compared with 107 in 1982. By the mid 1990s, the total birth rate in China was 1.8 births per woman and the population in 2000 was 1.27 billion demonstrating overall success of the policy, though in the context of some resistance and dramatic impacts on the personal and family lives of Chinese people from all sections of society. There is also growing concern about how China will support an aging population, with fewer people of working age supporting a growing number of retirees (White, 2003).

Donella and Dennis Meadows at the Massachusetts Institute for Technology and forecast material limits to economic growth, as well as developing alternative scenarios to avoid catastrophic collapse of renewable resources. The modelling was revisited in 2004 and whilst some of the more spectacular predictions for ecological collapse and the over abstraction of resources have not come to pass, other trends in resources and pollution are worse than originally forecast (Meadows et al., 2005).

Limits to Growth challenged the idea of unlimited economic growth, particularly given the strong connection between growth in economic activity and growth in resource extraction. The economic implications of environmental limits to growth were developed by ecological economists such as Herman Daly, Robert Costanza, Kenneth Boulding and Nicholas Georgescu-Roegen. Daly's book *Steady-State Economics* was published in 1977 bringing physical and ecological laws to bear on the physical economy and proposing the steady state economy as an alternative to the dominant model which is based on continual economic growth. The idea of the steady state economy is that there are physical and ecological limits to the expansion of economic activity based on material extraction and consumption. Ecological economists hold that economic activity will ultimately be constrained by physical and ecological limits, and it is preferable to develop an economic system that achieves steady state before breaching these limits.

1.3   INTERNATIONAL DEVELOPMENT

The idea of limits to growth in order to protect ecosystems stood in contrast to development models based on industrialisation, modernisation and economic growth. The response to reconstruction of Europe after the Second World War and the development of post-colonial nations in Africa, Asia and Latin America was to invest in industrialisation of the economy, including agricultural systems and the exploitation of natural resources such as timber, oil and minerals. In the 1960s and 70s, institutions that had been established for the reconstruction of Europe, such as the World Bank, the International Monetary Fund and the precursor to the World Trade Organisation, helped drive a model of development based on investment in large scale infrastructure, industrial development and international trade. Economic development was seen as the basis for human development. As economies were modernised and industrialised, they would build the necessary wealth to provide education and health care for their populations, reduce poverty, and implement environmental protection and pollution control measures. Much economic development was based on utilising technologies and systems from the developed world and involved foreign investment and expertise. During the nineteen eighties and nineties development, models focussed increasingly on the role of markets and the private sector in delivering infrastructure and industrial development, rather than state ownership and protection of industry through tariffs and subsidies.

The model of development based on industrialisation and exploitation of natural resources has delivered significant benefits to millions of people in developing countries. However, it has had some significant shortcomings and continues to struggle to deliver improvements in the lives of the world's poorest people. Attention to industrialisation without adequate environmental protection has led to widespread ecological degradation in many countries.

Critics of the dominant development models and proposals for alternatives emerged in the nineteen sixties and seventies. The British economist E. F. Schumacher's 1972 book *Small is Beautiful* criticised development models based on the transposition of large scale industries such as mining and manufacturing from the developed to developing countries (Schumacher, 1974). He argued that such industrial development was not only unlikely to succeed in alleviating poverty, but it was dehumanising and ecologically destructive. As the foundation for the appropriate and intermediate technology movement, his work argued for development based on technologies that were appropriate to the context in which they operated. This meant that technologies should be chosen, built, operated and maintained by local communities using locally available resources. Such technologies and systems were likely to be of a smaller scale than the dominant industrial models, but they were more likely to deliver the benefits of development to poor people.

Rural development models based on the industrialisation of agriculture and the implementation of green revolution technologies and techniques were also criticised as failing to account for local knowledge and conditions, failing to deliver benefits to the poorest members of rural communities, damaging local ecosystems and degrading soil and water. Alternative models of rural development emerged that were based on assessing community needs and incorporating the local knowledge of farmers as the starting point, rather beginning with new technologies and practices and rolling them out across rural communities (Chambers et al., 1989). This requires a much higher level of participation by local communities in the process of development than models based on government and private sector promotion of particular agricultural techniques or other modern technologies. Such participatory techniques require attention to local levels of education, gender and power relationships within communities and access to health and other basic services.

In recent decades, increasing attention has been paid to urban planning and development as more than half the world's population now live in cities (UN Habitat, 2006). The fastest growing urban regions are in developing countries, posing particular challenges for provision of infrastructure, housing, education, health, transport and other basic services. Most the growth in urban population occurs in slums, which are informal settlements that most often lack access to basic infrastructure and fall outside formal government planning processes. Urban development based on delivering conventional infrastructure often fails to meet the needs of the urban poor, delivering modern, world-class services to middle class and wealthy communities, but failing to address the complexities of rapid population growth, poverty and urban expansion (Marvin and Graham, 2001).

1.4   SUSTAINABLE DEVELOPMENT

Sustainable development emerged as a way through conflict and tradeoffs between environmental protection, economic development and meeting the needs of the world's poorest people. In its weakest form, it is a compromise which maintains dominant models of development by paying greater attention to environmental concerns, and in its strongest form, it proposes a radically different approach to economics, development and human relationships with the natural world (Baker, 2006, Neumayer, 2010). Sustainable development has been formalised through international institutions negotiations, with the United Nations playing a leading role through a number of important conferences, conventions and commissions.

The United Nations Conference on the Human Environment held in Stockholm in 1972 was the first in a series of international meetings which addressed the need to protect the environment whilst maintaining essential processes of development. The 1987 *Our Common Future* report by the World Commission on Environment and Development, also known as the Brundtland Report, comprehensively addressed the problem of sustainable development, which it defined as 'development which meets the needs of the present without compromising the ability of future generations to meet their own needs' (WCED, 1987). This definition demonstrates the importance of maintaining development in order to meet the needs of current and future generations, but highlights the need to manage resources and ecosystems for long term viability. Significantly, the Brundtland definition of sustainable development focuses on the needs of humans, without specifically mentioning obligation to protect ecosystems and other species for their own sake.

The Brundtland Report was followed by the United Nations Conference on Environment and Development in Rio de Janeiro in 1992. This conference was the largest ever gathering of international heads of state and represented a watershed in international efforts to address sustainable development and the need for international co-operation on environmental protection. It was also significant as a major gathering of non-governmental organisations (NGOs), signalling their importance in delivering sustainable development. The conference resulted in important United Nations conventions relating to biodiversity, climate change and desertification. The Rio meeting also gave rise to Agenda 21 which outlined how sustainable development should be pursued and implemented into the twenty-first century. Local Agenda 21 gave particular emphasis to the role of local government in addressing the challenges of sustainable development, reflected in the slogan 'think global, act local.'

The 2002 World Summit on Sustainable Development was held in Johannesburg, with a much lower political profile than the meeting in Rio ten years earlier. The intervening years demonstrated the complexity of translating high level agreements into legally enforceable instruments and on the ground action. The private sector and business groups were more visible in Johannesburg than at previous conferences, highlighting growing awareness of sustainable development amongst corporations.

Sustainable development was also an important consideration in the 2000 UN Millennium Declaration which committed the international community to meeting a series of development targets by 2015. The members of the United Nations agreed to a series of targets relating to health, education, poverty, gender equality and sustainable development relating to measurable improvements from a 1990 baseline.

Each of the seven MDGs is comprised of more specific targets, which for the goal to 'ensure environmental sustainability' include:

- Integrate the principles of sustainable development into country policies and programmes; reverse loss of environmental resources.

United Nation conventions

The United Nations has been an important institution in developing international agreements which form the basis of international law in relation to the environment. The UN is especially important because many environmental issues such as pollution do not respect national boundaries. Conventions are agreed through international diplomatic processes and meetings of representatives of nations. The agreements must then be passed through national government and legislative processes before coming into effect, so national domestic politics is often an important element of international negotiations. Some examples of UN Conventions and the year they were instigated:

- Law of the Sea (1982).

- Montreal Protocol on Substances that Deplete the Ozone Layer (1987).

- Basel Convention (1989).

- Climate Change (1992).

- Biodiversity (1992).

- Desertification (1992).

Although the initial framework for the conventions have been agreed, many of them remain to be implemented as the details of the conventions have not been finalised and they have not been ratified through sufficient national legislatures.

- Reduce by half the proportion of people without sustainable access to safe drinking water.

- Achieve significant improvement in lives of at least 100 million slum dwellers, by 2020.

The target relating to improving access to safe drinking water is one of the few that is on target to be achieved, demonstrating the value of international agreements and targets. However, progress on the MDGs has been mostly disappointing, with key targets remaining beyond the reach of many individual nations and the international community (UN, 2011).

Millenium Development Goals

1. Eradicate extreme poverty and hunger.

2. Achieve universal primary education.

3. Promote gender equality and empower women.

4. Reduce child mortality.

5. Improve maternal health.

6. Combat HIV/AIDS, malaria and other diseases.

7. Ensure environmental sustainability.

8. Develop a global partnership for development.

1.5   DEFINING SUSTAINABLE DEVELOPMENT

Sustainable development is a contested concept. The Brundtland definition has come to be the standard, but it has been criticised for underemphasising the importance of environmental protection. Alternative definitions of sustainable development have been elaborated by governments, non-governmental organisations, scholars and professional institutions.

A common representation of sustainable development is as the intersection of economic, environmental and social considerations in development (Figure 1.2).

An alternative representation, preferred by ecological economists, is to represent sustainable development concentric circles, emphasising the dependence of the economy on social and ecological systems (Figure 1.3).

Sustainable development has been translated into different sectors and political contexts. Each country has taken a slightly different approach to implementing sustainability in government and businesses, non-government organisations and communities have also been important in interpreting and implementing sustainable development. Sustainability consulting is a growing business area for many engineers and other professionals, particularly in developed economies.

Pursuing sustainability usually involves acknowledging the importance of ecological, social and economic factors in determining the success of development. As a starting point, these factors are considered in parallel and tradeoffs are acknowledged. As understanding of sustainability grows,

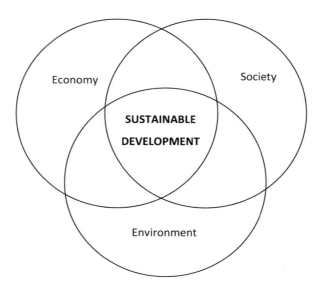

Figure 1.2: Intersecting circles of sustainable development.

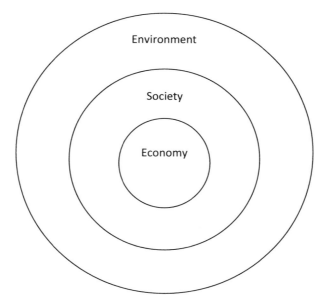

Figure 1.3: Concentric circles of ecological economics.

opportunities are created for integrating these three different domains of concern, looking for synergism and mutual gain. Achieving sustainability involves respecting ecological limits to the growth in human demand for resources and the capacity of the environment to absorb pollution, recognising the value of development in delivering improved quality of life, particularly to the world's poorest, and acknowledging the importance of local communities and individual citizens in achieving and maintaining change.

CHAPTER 2

Ecological Modernisation

Sustainable development requires consideration of social, ecological and economic factors in decision-making, design and management of development. Whilst this is widely recognised and agreed in principle, it has proved difficult to achieve in policy and practice. A more common response to the ecological crisis, particularly in developed countries, has been to try to incorporate environmental protection and resource conservation into existing institutions of government, industry and society. In Europe and elsewhere, this has been characterised as 'ecological modernisation' – a policy response which maintains the existing structures of the economy and society and reforms them to incorporate the need to protect and restore the environment. Ecological modernisation provides an important policy framework for many of the efforts of engineers to contribute to sustainable development. The development of new technologies and systems for managing the environmental impacts of industrial production are central to the ecological modernisation agenda, and engineers have a clear role to play in achieving this.

Engineers have been central to the development of the modern industrial economy and so have a vital role to play in ecological modernisation, as this model of progress adjusts in response to the global ecological crisis. If engineers are to further contribute to building sustainable societies, it is helpful to analyse prominent engineering methods and responses in terms of ecological modernisation policies and techniques. This helps to explain the important role that engineers have to play in the greening of industrialisation, but it also provides some insights as to why engineering contributions to sustainable development have been largely limited to environmental protection and resource efficiency, with limited incorporation of social considerations.

This chapter describes the key features of ecological modernisation and it implications for engineering practice. It begins by considering the nature of modern society and institutions, analyses ecological modernisation as a policy response to the global ecological crisis, and outlines key tools and techniques used by engineers to develop new systems and technologies to deliver a greener economy. The limitations of ecological modernisation in achieving the goals of sustainable development are considered, providing avenues for further development of alternative models of engineering practice which are better able to address social as well as ecological and economic concerns.

2.1 MODERNISATION

Modernisation is central to engineering. It implies processes of continual progress, involving the development of new ideas, knowledge, technologies and systems. Engineers contribute to the modernisation of society by developing new technologies and systems. Not only are engineered technologies and systems constantly developing, but engineers also innovate and modernise their own methods for delivering new technologies, infrastructure and services to society. Engineering is an essentially modern profession.

Ideas about modernisation, modernity and the nature of modern society are central to social science and theory. Modernity refers to the most recent historical epoch, and it can be traced to the industrial revolution and the dramatic social and political changes that occurred during that period. Modernity can be characterised by the emergence of institutions and structures which are now largely taken for granted, such as the nation state, the military, capitalism, the professions and science. Modernity can be characterised by a belief in the benefits of progress in science, technology and rationality. Modern societies contrast with traditional societies in their focus on the future and prospects for improvement based on ever expanding knowledge, rather than reverence of the past and maintaining long held practices and beliefs.

The period since the 1960s has presented challenges to modernisation and the structures of modern society. The ecological crisis, religious extremism, technological failures and controversies about new scientific developments have called into question the belief in the inevitable benefits of progress. Post-modern ideas in the arts, philosophy and social theory became influential in the nineteen eighties and nineties characterised by an abandonment of the idea of a unifying story of modernity and progress and the exploration of multiple, alternative narratives and representations of culture, self and reality.

2.2   ECOLOGICAL MODERNISATION

The ecological crisis, in particular, poses a significant challenge to modern institutions and an unquestioning faith in the benefits of scientific and technical progress. *The Silent Spring* was one of the first efforts to highlight significant and unanticipated consequences of technological development (Carson 1962). The impact of pesticides on wildlife and human health was not anticipated and potentially outweighed the benefits of these novel chemicals in improved crop production and public health. Scientific progress and industrialisation were shown to have significant, and possibly unacceptable, costs as well as benefits to society. Increasing knowledge about the impacts of industrialisation on the natural environment and the growing environmental movement led to considerable critique of the fundamental assumptions underpinning modern faith in science, technology and industrialisation. Proponents of the steady-state economy challenged fundamental capitalist assumptions about the necessity for economic growth. Environmental philosophers and activists posed fundamental questions about the structure of relationships between humans and the natural world, and within human society, proposing that the very structure of modern society and economies lay at the heart of the ecological crisis, as discussed in Chapter 3.

Against the background of radical critiques and proposals for revolutionary change to social, economic and political structures, governments and businesses made pragmatic responses to the growing problems of environmental degradation and resource depletion. A more reformist agenda emerged, which has been characterised as ecological modernisation. The essential assumption of ecological modernisation is that the structures of modern society and the economy are fundamentally sound, but need urgent reform to adequately account for ecological systems and constraints (Huber 2000, Huber 2004, Mol 1995, Mol and Sonnenfeld 2000). Technological innovation is at the heart of ecological modernisation reforms, delivering the possibility of continued economic development without increasing extraction of natural resources or pollution of the environment. Ecological modernisation policies aim to decouple economic and social development, and environmental harm.

Ecological modernisation requires a new role for science and technology in solving environmental problems and radically improving the efficiency of industrial processes. It has been associated with policies that propose an increased importance for market forces and the decisions of individual economic agents in choosing the most efficient and ecologically sound options for development. The environmental and other social movements have an important role to play in ecological modernist policies and processes, with much greater emphasis on partnership than opposition. This requires new ways of thinking and talking about environmental issues in society, politics, science and technology (Mol and Sonnenfeld 2000).

The role of the state in environmental protection has also changed under ecological modernisation. Early government responses to environmental problems focussed on regulating and controlling industrial activities to minimise environmental harm, while more recent ecological modernist approaches have seen the state move into co-ordinating and facilitating role, enabling the private sector and market forces to drive ecological innovation. Private sector engineers have a very important role to play in designing and delivering technological innovation within emerging markets for environmental services and to meet requirements for greater resource efficiency. Early ecological modernisation focuses largely on supply side problems, aiming to improve the industrial processes that support modern life, rather than requiring consumers and citizens to make significant changes to their lifestyles and behaviour (Barry 2005). More recently, ecological modernisation has incorporated demand management and consumer behaviour change as important policy objectives.

Ecological modernist policies promote the role of the market and private sector in improving ecological efficiency and environmental protection. Some examples include:

- Emissions trading schemes to efficiently achieve reductions in pollution.

- Environmental taxation to provide market signals that reflect environmental costs of products and services.

- Subsidies for development of ecological technologies such as renewable energy.

- Funding research and development in environmental technologies and systems.

- Privatisation of waste management and other environmental services.

Ecological modernisation is a theoretical framework that has encouraged development of a range of economic, policy, technical and management tools and techniques. These include environmental economics, environmental impact assessment, cleaner production, life cycle assessment, industrial ecology, ecological design and natural capitalism.

2.3   ENVIRONMENTAL ECONOMICS

Environmental economics is central to ecological modernisation. Environmental economics aims to incorporate the costs of environmental harm and depletion of natural resources into market prices and to devise new markets for environmental goods and services. In contrast to ecological economists, who aim to devise alternative economic systems that are based on ideas of steady state and limits to growth, environmental economists use and reform existing market based, capitalist economics to solve environmental problems. One fundamental aim is calculate the full economic value of environmental resources and to develop mechanisms for incorporating it into market signals. Another aim is to develop new markets, such as for carbon emissions, as a way of improging environmental performance using the most economically efficient means.

Environmental economists use a range of tools developed for conventional economic analysis and extend them to incorporate environmental factors and demonstrate the relationships between the health of the environment and the state of the economy. General equilibrium modelling of the macroeconomic policy decision has been used to demonstrate the importance of environmental goods such as wildlife to the economic performance of countries dependent on eco-tourism (Markandya *et al.* 2002). Contingency pricing techniques such as willingness-to-pay, which for instance may involve asking a random sample of people how much they would pay to preserve a wilderness area, have been used in cost-benefit calculations as part of environmental policy and impact assessment decisions.

2.4   ENVIRONMENTAL IMPACT ASSESSMENT

Environmental Impact Assessment (EIA) was one of the first policy measures implemented by governments around the world to address the need to incorporate environmental protection into development. The National Environmental Policy Act 1969 (NEPA) in the United States required, for the first time, full assessment of the environmental impacts of proposals for development involving federal agencies, laying the foundations for Environmental Impact Assessment (Bailey, 1997). Environmental Impact Assessment procedures were soon implemented in many industrialised countries including Australia, Canada, the Netherlands and Japan, and later the European Community (Wathern, 1988). Developing countries including Colombia, the Philippines and Thailand also implemented EIA procedures, although the general uptake was much slower outside industrialised economies (Ebisemiju, 1993; Sankoh, 1996).

EIA requires developers to systematically identify and quantify potential impacts using recognised scientific techniques. This requires a good understanding of the local environmental conditions before the development, identifying potentially sensitive or threatened species and ecological communities that could be impacted, and modelling of emissions and other impacts during the lifetime of the facility. Developers must demonstrate evaluation of alternatives and outline mitigation measures to minimise the impact of their proposal. EIA provides opportunities for developers to improve the performance of their proposal and provides planners and decision makers with information regarding the potential environmental risks and costs associated with a development before it goes ahead.

Planning procedures have evolved further in the last decade to move beyond EIA for individual development proposals to Strategic Environmental Assessment (SEA) which covers larger scale plans and processes and aims at a more integrated approach. SEA allows for greater public participation and requires a more comprehensive assessment and evaluation of alternatives.

2.5   CLEANER PRODUCTION

Cleaner production techniques emerged in the 1990s as a means for improving the environmental performance of manufacturing processes while at the same time saving money. Cleaner production fundamentally aims to move from 'end of pipe' technologies for environmental protection such as pollution control and waste disposal, to eliminate waste and pollution from upstream production processes. A core principle is that waste is not only an environmental concern, but it also costs businesses money in terms of wasted materials and disposal. By reducing wasted materials and energy, process improvements can also increase profitability. Cleaner production systems usually begin with an audit of water, energy and materials in the process, then involves staff in identifying opportunities for improvements, analysing the environmental and economic benefits of proposed changes and implementing those with the biggest payback.

2.6   LIFE CYCLE ASSESSMENT

Assessing the impacts of products on the environment is an important first step in identifying opportunities for improving efficiency and minimising waste and pollution. Life Cycle Assessment (LCA) techniques have developed since the 1990s which provide quantitative measures of the environmental impact of specific products from the extraction of raw materials, through to manufacturing, transport, packaging, use and disposal. LCA techniques allow for comparison of products and production processes and help to prioritise actions for improving environmental performance of products.

LCA principles have also been applied to buildings to help improve building design and to enable governments, clients and users of the building to compare the performance of different designs and buildings. Building LCA considers the environmental impacts of the materials used in construction and the energy and resources consumed throughout the life of the building. Several systems for analysis have been developed including the US based Leadership in Energy and Environmental Design (LEED), the UK based Building Research Establishment Environmental Assessment Methodology (BREEAM) and the Australian Building Sustainability Index (BASIX). These systems assess the environmental performance of building designs and provide ratings that are comparable between buildings. They cover different aspects of the design including water and energy efficiency, air quality, construction materials, building management systems, provision for sustainable transport options, waste management, and impacts on local land use and ecology. They are useful in setting standards across the construction and real estate industries and provide for flexibility to allow designers to innovate improving environmental performance.

2.7   INDUSTRIAL ECOLOGY

Industrial ecology aims to design and manage entire industries, industrial parks and regions using ecological principles (Huber 2000, Allenby 1999). Connections between industries are developed to maximise resources and energy efficiency. Thinking of industrial systems as ecosystems provides the basis for 'closing the loop', as waste energy or materials from one production system provide input to another. Industrial ecology requires co-operation between different companies and a level of transparency about their production processes and materials volumes. It can begin with relatively simple acts of co-operation such as agreeing to a common waste management contractor and system for waste collection and handling, and develop into ongoing contracts for material or energy transfer between companies. Municipal waste to energy and the use of fly ash from coal-fired power stations in brick and concrete manufacturing are well established processes that can further provide starting points for establishing ecological relationships between industries. Systems analysis can help to identify opportunities for industrial ecosystems and highlight the benefits to participants. Industrial ecosystems are often facilitated by local regulators but require strong leaderships and participation from all actors to succeed.

2.8   ECOLOGICAL DESIGN

Ecological design is increasingly important in developing new products, services, buildings and infrastructure. Good ecological design recognises natural systems as providing inspiration for innovation as well as constraints on the form and function of objects and buildings. Ecological design aims to minimise the impacts of products, structures and systems over their lifecycle.

Biomimicry refers to the use of natural systems as inspiration for design (Benyus 2002). Ecological systems have evolved to reuse resources and to use energy efficiently. Natural structures and materials provide can provide designers with useful ideas for how to achieve functional and aesthetic objectives whilst minimising ecological impacts. Ecological design is not only delivering new products, services and structures, but also new approaches to design and innovation that recognise opportunities as well as constraints in meeting environmental performance targets.

In 1969, landscape architect Ian McHarg (1992) published the book *Design with Nature*, which presented an approach to urban planning and landscape design that started from an in-depth understanding of the ecological, geological, hydrological and other natural processes that shaped places. Working *with* nature in this way was contrast to dominant approaches to urban planning, design and engineering which aimed to control and resist natural forces using hard engineering techniques to dramatically reshape landscapes to meet human visions or to accommodate unplanned growth. Using a series of case studies, McHarg proposed that using scientific understandings of biophysical processes to produce maps and criteria for development and land use not only allowed for preservation of natural places but also improved the health of urban areas and protected them from natural disasters such as storms and floods.

Ecological designers extend conventional concerns with eco-efficiency and waste minimisation to incorporate deeper ecological principles in design. William McDonough and Michael Braungart (2002) outline the principles of *Cradle to Cradle* design in their 2002 book, reconsidering materials as either biological nutrients which can be returned to biological ecosystems or industrial nutrients which can be reused and 'up-cycled' in future manufacturing or construction. Cradle-to-Cradle design incorporates detailed analysis of materials with these two classes of nutrients in mind, eliminates potentially toxic substances, and considers the multiple uses of the product, building or material over its lifetime, including recovery for future use in industrial or natural ecosystems.

Sim Van Der Ryn and Stuart Cowan (2007) define ecological design as 'any form of design that minimizes environmentally destructive impacts by integrating itself with living processes' (p.33). They bring together developments in ecological engineering, agriculture, architecture, urban design, landscape architecture, product design and environmental philosophy to draw out five common principles for ecological design:

1. Solutions grow from place – design must be locally specific and respect local ecological conditions

2. Ecological accounting informs design – scientific analysis of the materials and energy flows of natural systems as well as design propositions help to minimise environmental impacts and maximise benefits

3. Design with nature – natural systems provide inspiration for design as well as constraints

4. Everyone is a designer – participation by users, stakeholders and the community is important in ensuring good design and can promote innovation

5. Make nature visible – humans have an innate need for contact with nature and designs which make natural systems and processes visible help people to behave in ways that reduce their environmental impacts and improve their wellbeing.

2.9   NATURAL CAPITALISM

Natural Capitalism was the title of a 1999 book by Paul Hawken, Amory Lovins and Hunter Lovins. It makes the proposal for a fourth category of capital, natural capital, to be considered in market capitalism alongside the conventional financial capital, human capital and manufactured capital. Using the language of capitalism, they point out that industrialisation has so far drawn down natural capital, and that industrial systems should be radically transformed to reverse this trend. Industrial systems as a starting point should only draw on the 'interest' provided from natural systems, not the 'capital' which produces essential natural services and resources. Furthermore, industrial systems should restore ecosystems to recover the balance of natural capital that has been lost since the industrial revolution. Natural capitalism is based on four key principles: radical resource productivity, biomimicry, service and flow economy, and investing in natural capital.

Radical resource productivity is essential to make the best use of resources that are extracted from the environment. A ten-fold improvement in the efficiency of resource use is required and achievable according to the principles of natural capitalism. Redesigning products and manufacturing systems can lead to a ten-fold improvement in resource productivity by reducing waste, maximising recycling, improving durability, minimising material requirements and other actions.

Natural capitalism recognises the value of biomimicry in design of products, services and systems. The innate efficiency of natural systems provides inspiration for industrial ecology, ecological design, and the wider arrangement of social and economic systems. New products and systems inspired by nature can be found in fields as diverse as architecture, agriculture, electronics, drug design and materials engineering (Benyus 2002).

The service and flow economy proposes to decouple economic growth from material and energy consumption. In a service and flow economy, consumers purchase the service that the product provides, rather than the product itself. The service provider is then able to deliver resource efficiency and life-cycle management of products in a way that individual owners of products cannot. For instance, if consumers paid for their homes to be kept at a constant temperature, rather than purchasing a boiler, air conditioner and the fuel and electricity to run them, 'thermal service providers' could analyse the needs of the homeowner, and as required could insulate their house and install and maintain the most efficient technology for temperature control. Rather than each home owning a washing machine that is only used for a few hours each week and is rarely maintained or upgraded, if consumers paid for laundry services, clothes could be laundered using the most energy and water efficient machines available, owned by the service provider and utilised full time. A service and flow economy exploits economic opportunities in managing flows of materials and energy to improve efficiency, rather than a resource based economy where economic value is primary generated by the extraction and consumption of resources.

Investing in natural capital is required to restore ecosystems that have been lost. Replenishing the stocks of natural capital increases the long term sustainability and viability of social and economic systems that rely on ecosystem services and resources. Natural capitalism recognises that the challenge of reforming industrial systems is not only to slow down current rates of ecosystem degradation but to reverse these processes so that natural systems can be restored to provide services and resources for future generations.

2.10   TECHNOLOGICAL OPTIMISM

Ecological modernisation is at once highly optimistic and essentially pragmatic. It recognises the strength of modern institutions, industrial systems and society and aims to modify the processes of modernisation rather than undermine them in order to protect and restore the environment. It maintains an implicit assumption that modernisation and industrialisation will lead to economic and social development, which is ultimately in the interest of all members of society, including the poor.

Ecological modernisation policies were first noted in the UK and Netherlands (Hajer 1997), but they have come to characterise many international and global approaches to addressing environmental problems. For instance, international climate change negotiations are dependent upon assumptions about whether or not reductions in emissions are possible without harming economic growth. At the level of international negotiations, there is consensus that action should be taken to avoid catastrophic climate change, but not if it requires slowing economic growth. Technologies such as nuclear and renewable energy, and policy measures such as emissions trading schemes, are all proposed as means by which the processes of modernisation can be maintained around the world without increasing carbon emissions. Debate abounds as to whether or not carbon reductions can be achieved without detrimental economic impacts, not about whether or not the economy should continue to grow or about alternatives to modernisation.

Ecological modernisation policies optimism in the capacity of new technologies to solve environmental problems risks underestimating the problem of social acceptance of new technologies. Whilst engineers and industrial managers may be capable of developing new technologies and systems that improve resource efficiency or minimise pollution, their success depends on being acceptable and adopted by users and society. Nuclear energy remains a controversial technology despite its potential to reduce carbon emissions. Recycling of municipal wastewater to provide a new water resource has proved to be unacceptable to residents of cities in Australia and US, despite technical assurance of low health risks (Bell and Aitken 2008).

Ecological modernisation does not directly address problems of uneven distribution of costs and benefits of new technologies and systems. The progress towards a service economy and the decline of polluting manufacturing industries in Europe has improved environmental conditions there, but at the cost of environmental pollution in countries in Asia and Africa, where the polluting industries have relocated to. European citizens enjoy a cleaner environment whilst consuming products imported from other countries, which may be increasingly polluted.

Such challenges to the implementation of new technology reflect wider public concerns about the risks of modernisation, which reflect deeper questioning about industrialisation and technological progress that could be seen in the early environmental movement. Ecological modernisation may be the most pragmatic response by policy makers and industrialists to the ecological crisis, but it avoids deeper questioning about the structure of society and cultures that drive continual growth in consumption of resources, maintain deep inequalities between rich and poor and tolerate the destruction of essential ecosystem services and natural places.

CHAPTER 3

Environmental Ethics

From the earliest years of the environmental movement, scholars and activists have tried to understand why and how humanity is capable of ecological destruction on such a vast scale. Radical ecologists have not been satisfied with the ecological modernists' pragmatic account and have looked for deeper cultural origins for the ecological crisis. Several distinct movements have developed within environmental philosophy and activism, each with a particular analysis and position on the underlying reasons for humanity's destruction of nature and propositions for alternative social and political structures. Knowledge of the different elements within environmental philosophy is useful in itself, for their analysis and propositions, and also in order to understand the complexity of the environmental movement and environmental politics. Donna Riley (2008) in her volume *Engineering and Social Justice* identifies ecological justice as an issue that is central to engineering practice and efforts within the profession to make a positive contribution to social justice and equity.

Ecological modernisation is a pragmatic, but also highly optimistic response to the ecological crisis. Whilst some advances have been made in reducing pollution and improving resource management in particular countries, the environment has continued to be degraded at an alarming rate since the environmental crisis emerged in the 1960s. The processes of industrialisation proceed with such force that ambitions for radical resource efficiency, restorative environmental technologies and industrial ecology remain largely unfulfilled decades after they were first proposed. Global economic growth has not been decoupled from growth in resource use and pollution, and the rapid industrialisation of China, India, Brazil and other developing countries is largely following the conventional pathway of resource exploitation and rising pollution, despite significant environmental improvements. The environment and sustainable development are now more important to engineering than ever before, and a growing number of engineers now work on environmental technologies and designs. However, the vast majority of the engineering work, research and teaching remain largely unconcerned about the environmental crisis and problems of sustainability.

Given the scale and complexity of sustainable development and the ecological crisis, it seems worthwhile contemplating deeper analysis of the relationships between humans and nature. This chapter will describe some early efforts at environmental philosophy and scholarship before describing three key movements in detail – deep ecology, social ecology and ecological feminism. I will then provide a brief overview of some of the key elements of environmental politics and its influence on mainstream political agendas. The chapter concludes by considering the implications of radical critiques of modernisation for engineering and sustainable development.

3.1 ORIGINS OF ENVIRONMENTAL PHILOSOPHY

Philosophers, historians and social scientists efforts to understand the underlying reasons for the ecological crisis started alongside early scientific analysis in the 1960s. In 1968, the journal *Science* published an article by historian Lynn White titled 'The historical roots of our ecological crisis.' White argued that the ecological crisis could be attributed to the Judeo-Christian worldview and biblical origin stories which gave man dominion over nature. The positioning of humans as masters over the natural world and the belief that nature had been provided for use and exploitation by humans formed the basis of an essentially destructive relationship with the environment. The belief that man was made in God's image whilst the rest of creation was soulless further contributed to a culture of exploitation and destruction towards nature. His analysis of biblical depictions of the relationship between humans and nature has since been challenged by theologians who point to the importance of human stewardship of creation, rather than domination, within the Judaeo-Christian tradition. However, Lynn White's contribution was highly significant as one of the first mainstream analyses of the cultural origins of the environmental crisis, and his analysis of the domination of nature by humans is a theme that has been further developed in most branches of environmental philosophy.

Although not as high profile when first published, *The Sand County Almanac* by Aldo Leopold (1970) became one of the foundational texts for environmental philosophy. First published in 1949, *The Sand County Almanac* was a collection of Leopold's observations of the natural cycle of seasons on his small farm in Sand County, Wisconsin and reflections on the ethical responsibilities of land ownership. The book of essays culminated in his formulation of a 'land ethic' which stated that 'A thing is right when it tends to preserve the integrity, stability, and beauty of the biotic community. It is wrong when it tends otherwise.' Leopold's land ethic was an attempt to go beyond the economic rationale for land conservation that was being promoted in the US following the disastrous dust bowl of the 1930s, and as such provided an inspiration for later environmental philosophers looking to develop broader frameworks for environmental ethics.

Environmental philosophy has developed within many different schools of thought – some essentially novel, others incorporating environmental concerns into existing analytical, theoretical or political frameworks. Environmental thought has developed in diverse directions including liberal environmentalism, ecotheology, preservationism, animal rights, Marxist ecology and radical environmentalism. Environmental philosophy, politics and ethics seek to understand human relationships with nature, which often holds parallels and connections with how humans relate to each other and organise society. Key questions arise about the ethics of human interactions with non-human nature and how to best arrange human societies in order to achieve an environmentally ethical and sustainable future. Deep ecology, social ecology and ecological feminism are three prominent schools of thought in environmental philosophy, politics and activism, each concerned with how to arrange societies in order to avoid human domination and destruction of nature.

3.2 DEEP ECOLOGY

Deep ecologists hold that the *anthropocentrism* (human centred view) of dominant western value systems is at the root of the ecological crisis. They call for an *ecocentric* system of values and society that places ecological concerns at the heart of all human culture and politics. Since human systems are part of ecological systems, our primary concern should be the preservation and maintenance of nature. Moreover, an ecocentric world view values nature *for its own sake*, in contrast to an anthropocentric view which values nature only in terms of its use for humans. Non-human nature, particularly in wilderness areas, has a right to exist on its own terms, irrespective of potential economic or other value to humans. Deep ecology is closely associated with wilderness preservation, is generally anti-industrialist and supports strict control of human populations.

The term 'deep ecology' was first coined in 1972 in a paper by Norwegian philosopher Arne Naess (1995a). He contrasted 'deep ecology,' based on a deep questioning of human relationships to nature, to 'shallow ecology' which characterises more conventional scientific and reformist approaches. Shallow ecology is anthropocentric, focuses primarily on pollution and resource depletion, and is ultimately concerned with the health and affluence of people in developed countries. Deep ecology involves a deep questioning of the goals and viability of industrial society, focuses on the interconnectedness of all life, and aims at restructuring society to achieve greater local autonomy and decentralisation.

Together with American George Sessions, Naess outlined an '8 point platform' for deep ecology in 1984 (Naess and Sessions, 1995b). These are:

1. The well-being and flourishing of human and non-human life on Earth have value in themselves (synonyms: intrinsic value, inherent worth). These values are independent of the usefulness of the non-human world for human purposes.

2. Richness and diversity of life forms contribute to the realization of these values and are also values in themselves.

3. Humans have no right to reduce this richness and diversity except to satisfy vital needs.

4. The flourishing of human life and cultures is compatible with a substantially smaller human population. The flourishing of non-human life *requires* a smaller human population.

5. Present human interference with the non-human world is excessive, and the situation is rapidly worsening.

6. Policies must therefore be changed. These policies affect basic economic, technological, and ideological structures. The resulting state of affairs will be deeply different from the present.

7. The ideological change will be mainly that of appreciating life quality (dwelling in situations of inherent value) rather than adhering to an increasingly higher standard of living. There will be a profound awareness of the difference between bigness and greatness.

8. Those who subscribe to the foregoing points have an obligation directly or indirectly to try to implement the necessary changes.

Propositions for restructuring economic, technological, ideological and social systems in line with the tenants of deep ecology centre on the concept of bioregionalism. Bioregionalism proposes smaller scale, decentred human settlements which live within the means of their local ecological systems (Hay, 2002). Bioregional development maintains large zones for 'free nature' to allow basic evolutionary processes to proceed without human interference. Buffer zones between free nature and human settlement would provide corridors for wildlife and conservation. Human settlements, including cities, would also provide space for wild nature and would use local resources to meet basic human needs.

Deep ecology has been associated with some of the most militant environmental activism, particularly associated with the wilderness conservation movement in the USA. Activists campaigning against roads in wilderness areas and logging of old growth, frontier forests claim to represent ecocentric values against the anthropocentrism of industrial development and resources exploitation. Protection of wilderness areas for their own sake has been a central element of conservation campaigns, which has met with some success internationally. However, the association of deep ecology with some of the most militant environmental activists is a point of controversy for the movement. Radical organisations such as *Earth First!* in the US have been associated with tactics such as sabotaging logging equipment and spiking trees to increase the risk of physical harm to chainsaw operators, demonstrating a disregard for human safety and welfare in defence of nature. *Earth First!* founder and deep ecologist Dave Foreman (1993) famously outlined such techniques and tactics in his book *Ecodefence: a field-guide to monkey wrenching*.

Deep ecology has been criticised by others in the environmental movement for a lack of attention to the structure of society and its relationship to dominance of nature. Its primary attention to the problems of anthropocentrism and its efforts to promote an ecocentric philosophy have been criticised for overlooking the relationships of domination and subjugation that occur within human society. Social ecologists and ecological feminists in particular have been critical of deep ecology, and their analysis and propositions for change draw on and extend existing critical frameworks to incorporate not only the oppression of nature but the oppression of the poor, women and others in society (Bookchin, 1988, Salleh, 1984).

3.3   SOCIAL ECOLOGY

The most prominent social ecologist of the twentieth century and clearest proponent of this branch of environmental philosophy was American Murray Bookchin. Bookchin's work, including his 1982 book *The Ecology of Freedom,* outlined key elements of social ecology which is based on the proposition that the ecological crisis arises from deep seated social problems due to the hierarchical structure of modern capitalist society (Bookchin, 1995). Human domination of nature is preceded by hierarchical relationships of domination and subjugation within industrial society. Social ecologists favour decentralised, local communities based on principles of self-organisation, which are in balance with local ecosystems as a means for re-organising society and resolving the ecological crisis.

The central ideas of social ecology and Bookchin's early work can be linked to anarchist philosophers from the late nineteenth and early twentieth centuries, such as Élisée Reclus and Peter Kropotkin (Clark, 1998, Hay, 2002). They were critical of the centralised state and authoritarian socialism, as well as capitalism, social Darwinism and other ideas based on an assumption that competition forms the basis of most productive social, economic and political interactions. Drawing on their professional expertise in geography and biology, anarchists such as Reclus and Kropotkin used examples from non-human species and pre-modern cultures of social grouping based on co-operation and self-organisation, rather than either competition or central control.

Social ecology also draws on the ideas of pioneering urban planner Patrick Geddes and philosopher of technology Lewis Mumford (Clark, 1998). Patrick Geddes was a Scottish biologist who applied ideas of biological organisation to urban planning, and promoted nature conservation. Geddes contributed to the emerging field of urban planning at the turn of the twentieth century highlighting the relationship between spatial and social organisation. His plans for cities involved decentralised communities living and working in harmony with local culture and environment, which would be enhanced by the introduction of new technologies. His work influenced fellow Scot, Ebenezer Howard, the founder of the idea of the 'Garden City' which became influential in mid-twentieth century urban design and planning, providing space within the city for industry, residences parks and food production, as the basis for a healthy, self-contained urban communities.

In contrast to deep ecology, social ecology places relationships of domination between humans at the centre of the ecological crisis. Drawing on long and varied traditions of localism, decentralisation and the use of ecological examples and metaphors in designing human settlements and social structures, social ecology has had a significant influence on environmental politics and activism. However, Bookchin's strong critiques of deep ecology and other elements of the environmental movement contributed to divisions within environmental politics and activism (Hay, 2002).

3.4   ECOLOGICAL FEMINISM

Ecological feminism (ecofeminism) identifies the cultural roots of the ecological crisis in the connection between the domination of women and the domination of nature. *Critical ecofeminism* uses scholarly and political analysis to show that the domination of women and nature have the same historical and philosophical basis, and proposes that solutions to the ecological crisis must address relationships of domination within the western cultural tradition, particularly gender relations. More controversially, *essentialist ecofeminism* draws direct parallels between women and nature, claiming that women are inherently closer to nature than men, and therefore suffer more directly when nature is dominated or destroyed. By extension, essentialist ecofeminists claim that women are more nurturing and better able to understand and care for nature than men. Essentialist ecofeminism is rejected by many feminists for whom the direct association of women and nature re-enforces patriarchal systems of thought and stereotypes of women as more emotional and less rational than men.

Karen Warren (1987) outlined four minimal claims of ecofeminism:

1. Connection between oppression of women and nature.

2. Linked oppression is sanctioned by a patriarchal framework.

3. Critique of patriarchy grounded in ecological principles.

4. Ecological politics must be feminist.

These demonstrate the need for ecological and feminist politics and action to be linked, as in fact they are both working to overcome the same structures of oppression. Although specific ecological and feminist issues may be distinct, they are linked through the same patterns of domination.

Historical analysis has been important in developing the core ideas of critical ecological feminism. Carolyn Merchant's (1990) *The Death of Nature* analysed connections between women and nature in language, culture and history before and after the scientific revolution. Prior to the 1500s, nature was described as an organism, both nurturing and uncontrollable, and clearly associated with the femininity. Since the scientific revolution, nature has been understood as a machine which can be controlled and dominated for the benefit of humans. Robert Tredgold's definition of civil engineering as 'the art of directing the great sources of power in nature for the use and convenience of man' is an example of this way of talking about nature. Metaphors of nature as nurturer have disappeared from dominant cultural discourse, but nature remains closely associated with women. Founder of the scientific revolution, Francis Bacon's volume *The Masculine Birth of Time* contains numerous examples of the domination and repression of nature using female metaphors and analogies, such as 'I am come in very truth leading you to nature with all her children to bind her to your service and make her your slave.'

Australian philosopher Val Plumwood (1993) extended the feminist concept of dualism to explain the philosophical structures of oppression that underpin the domination of both women and nature. A dualism is a conceptual pair, in which the two sides of the pair are defined as opposites in relation to one another by setting one as the superior and one as the inferior element. The definition of the superiority of the dominant side of the dualism depends on assigning inferior traits to the dominated element. Plumwood outlines a structure of domination whereby the dominant groups in western culture are defined in relation to their dualistic opposites, which are culturally constructed as inferior. For example, male and female are biologically different, but masculinity and femininity are culturally constructed as opposite genders, with masculine traits dominant in the structures of power in society. Masculinity is associated with power, which can only operate if there exists an opposite set of traits, femininity, that can be defined as weakness. In a dualistic relationship, the position of mastery by one category depends on the subordination of the 'other,' for example:

Masculine-Feminine.

Rational-Emotional.

Mind-Body.

Straight-Gay.

Master-Slave.

Racially White-Racially Black.

Modern-Tradition.

Culture-Nature.

Plumwood analyses the separation of nature and culture in western philosophy as a dualistic relationship. She traces this relationship through Greek and Enlightenment philosophy. The same dualistic structure which assigns characteristics of weakness to the feminine in order to define the power and dominance of the masculine, aligns nature with women and culture with men. Women are more associated with nature and bodily experiences such as caring for children and the sick, which is defined as inferior in relation to paid work associated with the mind and culture, such as the professions of law, medicine and engineering. Consequently, the definition of women and nature as inferior to men and culture sets up relationships of domination.

Strategies resolving the ecological crisis require disrupting dualistic structures of thought in western culture. These are the same dualistic structures which perpetuate the domination of women. Ecofeminist responses to the dualistic structure of domination are to redefine relationships between masculine and feminine, mind and body, nature and culture in terms of equality, empathy and mutual respect rather than domination. An alternative is to disrupt accepted categories to create new ways of analysing and talking about the world that undermine dualistic structures of domination. Recognising that the separation of nature and culture is a philosophical construct and the categories masculine and feminine are also socially agreed rather than biologically determined provide a starting point for reconsidering structures of thought based on domination.

Ecofeminist activism addresses a range of environmental issues, including urban, environmental justice and livelihood issues. Ecofeminists have been associated with nuclear disarmament and other anti-military activism. The Chipko movement in northern India is a well known ecofeminist movement which involved rural women literally hugging trees by holding hands around tree trunks as non-violent protest in opposition to logging and road building. They connected their livelihoods as women working on forest-based subsistence activities to the protection of the forests. Their slogan 'what do the forests bear? Soil, water and clean air' highlights the value of the forests beyond the immediate economic value of the timber and their connection to local livelihoods and wellbeing.

The debate between essentialist and critical ecofeminism has been a significant feature of the movement. Making direct, essential connections between women and nature through claims that women are fundamentally connected to nature, through childbirth, menstruation and essential characteristics of nurturing and care, in a way that men are not has been a popular idea in some activist movements, speaking directly to the experience of many women. However, it has undermined more powerful critiques which aim to disrupt the philosophical structures which create and enforce such associations. It also alienates mainstream feminists who see representations of women and nature as essentially connected as one of the elements of patriarchy which needs to be dismantled. Clarifying the distinction between sex, being the biological categories of male and female, and gender, which come from cultural definition of masculine and feminine, has been an important element of modern feminism. Essentialist ecofeminism risks associating the female biology, feminine characteristics of nurturing and care, and nature, undermining women's claims to be able to engage with public life and culture on the same terms as men.

3.5   ENVIRONMENTAL POLITICS

Since the 1970s, environmental activists have formed political parties and used democratic actions as a means of achieving change to address specific environmental issues as well as the general environmental crisis (Dobson, 2000). The first specifically environmental political party was the United Tasmania Group who endorsed candidates in the 1972 state elections in Tasmania on a platform of wilderness preservation in opposition to the construction of hydro-electric dams. The UK Ecology Party was formed in 1973, and green parties were soon established throughout the democratic world.

The tension between ecological modernisation and radical ecology has been prominent in debates within green political parties, particularly since the 1990s. Many green politicians have strong activist backgrounds supported by ecological philosophical frameworks such as deep ecology, social ecology and ecofeminism. However, the pragmatic realities of achieving change within existing democratic structures has led green politics towards ecological modernist policies and engagement with industries and corporations who are seen by many activists to be at the heart of the ecological crisis and other structures of oppression in society.

The pragmatic ecological modernist approach has also been associated with the greening of the mainstream political agenda. Environmental policy is now an established element of most political manifestos and government functions. Debate about environmental policy often reflects broader political debates and ideologies such as the role of markets versus regulation, or individual versus collective responsibility. Liberal politics promotes the importance of individual freedom and decision making, which can translate to a focus on the role of individual consumers and citizens in addressing environmental problems, for instance through education campaigns or green consumerism. Social democratic politics traditionally focuses on more collective responsibilities and action, and, in the environmental context, could promote a greater role for government in steering the economy and society through environmental taxes and regulations. Conservative politics traditionally promotes smaller government, particularly national governments, which can lead to decentralisation of decision making and service provision. Although coming from very different ideological positions, some of the policy outcomes of these different political agendas can be compatible with the goals of more radical ecological politics.

International efforts to develop agreements and conventions to resolve environmental issues such as the climate change or biodiversity are often undermined by domestic politics. Whether or not a national government agrees to an international treaty is dependent on their ability to translate the terms of that agreement into national law. National politics are important in this process and can undermine international efforts to address environmental issues.

3.6   SUSTAINABLE DEVELOPMENT

Seen through the framework of environmental ethics, mainstream sustainable development can be described as anthropocentric and based on shallow ecological principles. Rather than addressing fundamental relationships of domination between rich and poor, weak models of sustainable development maintain the conventional development imperative based on economic growth and modernisation (Baker, 2006, Neumayer, 2010). Sustainable development has been co-opted by dominant international power structures, thereby mitigating against necessary fundamental reform of society and the economy.

However, more radical formulations of sustainable development are possible. In addressing the needs of the poorest people and the environment, and addressing issues of concern such as preservation of wilderness areas and population growth, sustainable development can be understood as a means to address both social inequalities and ecological destruction. Radical responses to the ecological crisis require restructuring of society, which could be achieved by a strong version of sustainable development.

3.7   ENGINEERING

Engineering is a modern profession that has been implicated in vast ecological destruction and is central to implementing ecological modernisation policies. Engineering is the profession that focuses most clearly on technology, but engineering is not simply a technical profession. The technologies and systems which engineers develop reflect and reinforce particular social and economic structures (Baillie, 2006, Riley, 2008). Engineered systems have a strong impact on the behaviour of individuals and underpin expectations of what it means to live in a modern society.

Engineering ethics requires engineers to act to ensure the health and safety of the public and to work towards sustainable development. Acting in the best interests of the public and the environment increasingly requires engineers to reflect upon their role in modernisation and development. Whilst engineering is a pragmatic, client driven profession, in order for engineers to make a positive contribution to resolving global problems of ecology and poverty, and to maintain local ecosystems and cultures, they should be aware of relationships between society, politics and nature. Engineers should be aware of and accountable for their role in modernisation, ecological destruction and structures of domination.

3.8   MODERNITY

The separation of nature and culture is a central element of modern thought (Latour, 1993). It has enabled scientific investigation of nature and the development of modern industrial systems and technologies. Modern capitalist economies are based on continued economic growth, which is achieved through free markets. The domination and control of nature to achieve economic growth and improved standards of living is a central feature of modern society. Controlling nature can deliver unquestioned benefits to humanity, such as the elimination of the small pox virus or the development of systems for delivering clean water, but at same time has led to the destruction of biodiversity and created the risk of human induced climate change.

Radical ecological critiques of the domination of nature by modern industrial society highlight some deep questions about the nature of development and progress. Modernisation has clearly delivered tremendous benefits to millions of people around the world. However, there are limits to human control of nature. Relationships of domination which fail to recognise the limits of human control and absolute environmental limits to growth may ultimately endanger humans as well as destroying natural systems and harming other species.

The ecological crisis is a crisis of modernity. Ecological modernisation proposes that modernity can be reformed to address environmental problems and avoid the most severe forecast of environmental collapse. Alternatively, radical ecologists propose that the separation and domination of humans over nature is ethically problematic and reflects wider social power structures. Navigating a path between the pragmatism and optimism of ecological modernisation and the critical but idealistic analysis of radical ecology is a difficult task. Ecological modernisation maintains the basic separation of nature and culture, but it reforms institutions to avoid environmental harm, while radical ecology seeks to develop more holistic patterns of human development and settlement that are not based on the separation and domination of humans and nature.

Engineers' role in ecological modernisation is clear, focussed on the development of new technologies and systems to deliver ecological efficiency. Engineers' role in more radical or holistic alternatives to modern society is more problematic and requires deeper reflection on the role of the profession in facilitating human relationships to nature. However, if engineers are to contribute to stronger models of sustainable development which address social as well as ecological issues, it is important to reflect upon the nature of modernity and the relationship between technology, society and nature.

CHAPTER 4

Society and Technology

Technology, new and old, is of central importance to sustainable development. Existing technologies have facilitated the consumption of resources, destruction of ecosystems and pollution of the environment. New technologies hold the promise of more radical resources efficiency and ecosystem recovery. However, technology alone neither restores nor destroys ecosystems. Technologies are used by people, for good and bad purposes. The relationships between the nature of modern technologies, their use and their impact on society and the environment are the subject of much scholarship and debate. The impacts of technologies and systems on environment and society are of ethical importance to engineers, who are responsible for their design, implementation, operation and maintenance.

Discussions of sustainability can become polarised between belief in either technological fix or social change as the means to address environmental problems and alleviate poverty. Optimism in technology can be associated with ecological modernist policies and a general belief that modern social conventions cannot or should not be changed, and that technology will deliver sustainable outcomes without the need for social, behavioural or cultural change. A preference for social change as the means to resolving sustainability may be associated with more radical ecological perspectives, or simply a more circumspect view of the promise of technology to deliver the required changes in resource use and consumption.

Both social and technological changes are required to achieve sustainability. Indeed, social and technological changes often occur together. Mobile phones have changed the way we communicate and organise our personal and working lives. Rising wages and access to credit made cars more affordable and contributed to the development of transport systems based on individual car ownership in many modern cities. It is often impossible to meaningfully separate technological and social change. However, professional and academic expertise usually divides the world into technical and social knowledge, as if they were two completely separate domains. Engineers and scientists know about technology, and politicians, social scientists, historians and economists know about society.

In her volume in this series *Engineering and Society,* Caroline Baillie (2009) identifies three approaches which are associated with an interactive view of science and technology and society – the social construction of technology, technological systems, and actor-network theory. This chapter briefly introduces the idea of social construction before outlining the central elements of actor-network theory. Actor-network theory is proposed as a method for understanding how technology and society interact, which may be helpful in understanding how engineers can help build a more sustainable society.

In recent decades, social scientists and philosophers have shown increasing interest in analysing the social and political implications of science and technology. In the 1970s, social scientists began studying research laboratories using the methods of anthropology, to understand the social practices of scientists and engineers. This work showed that the creation of scientific facts and engineering artefacts are highly social processes, involving negotiation and agreement over facts, funding and resources, as much as the direct interrogation of nature to reveal new knowledge. It also highlighted the importance of scientific instruments, recording devices, tools, paper and other objects in the development, transmission and confirmation of scientific facts.

The new approach to studying science and technology was founded on a social constructivist understanding of reality and knowledge. Social constructivism can be contrast with the positivist view of reality and knowledge which underpins the scientific method. Positivism holds that reality is external and can be known directly. Science is the best way to know reality and involves experimentation, observation and objective analysis. Social constructivism holds that reality is what is agreed amongst social groups, and knowledge is created through social processes. Social constructivist research methods involve interpretation of different understandings of reality and the meanings individuals and groups ascribe to it, using qualitative methods including interviews, observation, and analysis of texts.

Constructivist approaches to technology highlight that the particular form of technologies is not the result of purely rational optimisation but also reflects social and political context and considerations. Bijker's (1997) analysis of the development of the bicycle is a classic study showing that the familiar form of the bicycle was not inevitable, and it became stabilised as safety considerations and demand for the bicycle as a form of transport won out over alternative designs which were faster and associated with cycling as sport. Both forms of the bicycle were manufactured and were viable alternatives, and the final form of the bicycle was not a technically rational decision but the outcome of social processes.

4.1   ACTOR-NETWORK THEORY

Actor-network theory is a branch of science and technology studies based on the work of Bruno Latour, Michel Callon and John Law. Their work began with anthropological studies of scientific laboratories and field sites. It was based on the simple proposition that the best way to understand what scientists and engineers do is to follow them around and ask them questions (Callon, 1986a,b, Callon et al., 1986, Latour, 1987, Latour and Woolgar, 1979). They also undertook historical studies of major developments in science and technology including the discovery of bacteria by Louis Pasteur and the development of navigation and sailing techniques that were essential to Portuguese exploration in the fifteenth century (Latour, 1998, Law, 1986).

Actor-network theory challenges the modern distinction between nature and culture, by pointing out all the 'hybrid' objects and phenomena that are neither natural nor cultural. As such, it provides a way of thinking through the problems of sustainability that avoid the problems dualism and domination that are highlighted by environmental theories such as social ecology and ecological feminism. It offers a new way of thinking about the interactions between technology and society that go beyond the technological optimism of ecological modernisation.

In his book *We have never been modern*, Bruno Latour (1993) identifies two dichotomies at work in modernity (Figure 4.1). The first is the dichotomy between nature and culture. This is the familiar distinction which is understood to underpin modern knowledge and society. The second dichotomy is between the work of purification and the work of translation. Purification is the efforts by modern society and knowledge to divide the world into the categories of nature and culture. Translation is the work of hybrid networks which link across the primary distinction between nature and culture. Although most representations of the modern world focus on the work of purification, which allows clear distinction between nature and culture, modernity relies on networks of human and non-human actors which cross over these categories. This is the role of hybrids, which are not as easily accounted for in modern systems of knowledge. Constructing and maintaining hybrid networks that cross between nature and culture could be considered as a definition of engineering. But this leads to a fundamental contradiction in how we think about modern culture in relation to nature. If to be modern means to live in a world divided between purely natural and purely cultural components, then as Latour's title suggests 'we have never been modern.'

Nature and culture exist only as categories constructed by modern knowledge systems, not as distinct realities. Modern knowledge systems have enabled the development of science, technology and society, including the many benefits of modern life as well as complex, hybrid problems that society must now address, such as sustainability. Solving these hybrid problems requires a way of understanding the world that does not begin by fundamentally dividing the world into nature or culture, technology or society. Understanding the world in terms of hybrid networks of human and non-human actors is the objective of actor-network theory.

4.2   STUDYING SCIENCE AND TECHNOLOGY

The principles of actor-network theory were established through early studies of scientists and engineers at work developing new knowledge and technologies. Michel Callon's (1986a) study of an unsuccessful attempt by Electicite De France (EDF) to launch an electric car onto the French market in the 1970s showed the importance of building strong relationships between human and non-human actors across a range of domains and distances.

In 1973, EDF responded to the oil crisis and rising environmental values amongst consumers with a proposal for an electric car. Their proposal was for car maker Renault to produce the car bodies and a third company to manufacture the electric engines based on the development of new electrochemical batteries. The proposal for an electric car also required particular arrangements in urban and transport planning, allowing for growth in car based transport and provision of points for recharging the cars' batteries.

An 'actor-network,' as defined by Callon, describes all the relationships between the different actors, such as EDF, Renault, the engineers developing the fuel cells, the catalysts and other materials that made up the fuel cells and engines, town planning official and car drivers. Where a set of

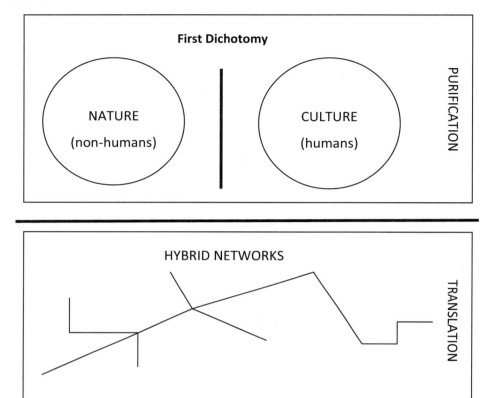

Figure 4.1: Modern dichotomies (Latour, 1993, p. 11).

relationships has become stabilised an actor-network can become a 'black-box,' where the internal relationships are no longer of importance and the overall stable functioning of the actor-network can be taken for granted. In this sense, a black-box becomes a discrete actor in an actor-network, and it becomes apparent that actors are themselves networks of relationships, which can be stabilised or destabilised. However, relationships between actors can become destabilised if conditions or the interests of actors change. Black-boxes can open up to reveal the relationships and networks that constitute them. Actors can dissent from network relationships if their interests are not being effectively translated in the network.

In EDF's actor-network, the fuel cell was a black-boxed actor, simply providing power to the engine to drive the electric car. However, the actor-network which formed the fuel cell broke down when the catalysts that are central to the creation of electricity showed a tendency to become poisoned, undermining its function. With the black-box of the fuel cell opening up to reveal a poorly functioning catalyst, EDF's actor-network started to break down.

EDF's simplification of Renault as a producer of car bodies in their actor-network was also challenged, leading to further breakdown of electric car actor-network. Renault rejected their role in the network by developing their own expertise in electrochemistry, their superior relationships with regulators and administrators of transport and automotive industries, and their knowledge of the needs and desires of consumers. The EDF electric car project ultimately failed as they were unable to stabilise the actor-network that they were attempting to construct. However, Renault went on to release an electric vehicle to the commercial rather than domestic market, using the more stable lead accumulator batteries, rather than the unproven fuel cell technology.

The EDF electric car case study is foundational in demonstrating how to analyse science and technology in terms that acknowledges the importance of social as well as physical relationships and entities. Creating new scientific knowledge or technologies is not simply about objective experimentation, data collection and analysis, nor is it merely about building social and political networks. In order to succeed new technologies and systems require engineers and scientists to pay attention to both the human and non-human actors in the networks they are building.

Actor-network analysis is also useful to demonstrate how particular technologies lock in specific social relationships and patterns of behaviour. Where technologies and systems have become stabilised, they can be dependent on their capacity to successfully define how humans will behave. In a very simple example, Bruno Latour (1991) shows how a large weight attached to a hotel key drives hotel guests towards returning the key to reception before leaving the building much more successfully than a small sign asking them to 'return your key.' The message 'return your key' is implicit in the cumbersome weight attached to the key. Speed bumps on road are more successful at changing driver behaviour than signs saying 'please slow down.' Whilst this kind of analysis seems obvious in these simple examples, it is important to recognise that technologies, systems and simple objects have 'baked in' messages to users that encourage certain patterns of behaviour and social relationships. As Latour concludes, 'technology is society made durable.'

Actor-network theory has been widely used as the basis for research into science, technology and society. The key principle for describing and analysing actor-networks is to 'follow the actor.' Actors can be human and non-human, and as the research follows the actors, they will uncover the relationships between actors that comprise the actor-network. Actors are defined in relationship to other actors, not in isolation. Importantly, the relationships between actors are based on physical, material realities, not on abstract or 'merely social' constructs or concepts.

Technologies and systems can have values and behaviours 'baked in' that may not be desirable or compatible with sustainability. Underground railway systems in cities such as London which require commuters to use stairs and escalators stabilise social values at the time that they were constructed – that people in wheelchairs or with other mobility constraints are not as important in the life of the city as able bodied travellers. Urban water systems which provide a continuous supply to our houses so that water is constantly running from our taps send the message that water is endless, contradicting public awareness campaigns about the need to save water. As engineers design new technologies and systems, it is important to be aware of the messages and values that they are 'baking in.' Design methods such as value-sensitive design and constructive technology assessment, which are discussed in Chapter 7, provide some ideas about how to achieve this in practice.

Achieving sustainability is neither a purely technical nor purely social task. Engineers and others working for sustainable development are engaged in building hybrid networks of actors that are both human and non-human. While existing relationships with non-human natural actors are based largely on domination and exploitation, building actor-networks for sustainability will require greater attention to be paid to the needs of non-human natural actors, so that they can be sustained in the longer term for their own intrinsic value as well as their usefulness to humans.

4.3 HYBRID DECISIONS

An outcome of the modern division between nature and culture has been that decision making systems have struggled to deal with environmental and technological controversies. Decisions about the environment, science and technology often cross the boundaries between fact and value. Science and engineering conventionally deal with facts, and politics deals with values. Political systems devised to address competing human interests have not been able to deal with the interests of the environment or to adequately consider the social implications of new technologies. Decisions about the environment or technology have conventionally been delegated to experts; however, this is problematic where expert opinion is divided or is not aligned with popular sentiments. For issues that cannot be easily divided into either matters of fact or questions of value, new arrangements for making decisions need to be made. This potentially shifts the role of scientists and engineers who are called to give expert opinions on matters of fact where both facts and values may be uncertain.

An actor-network approach to building relationships between human and non-human actors can provide the basis for new arrangements for decision making and define new roles for engineers and scientists. Engineers and scientists maintain their position as experts who are able to speak authoritatively about non-human nature and technologies, but they become more fully strongly engaged with political and social decision making processes.

In his book 'Politics of Nature', Latour (2004) outlines four stages in collective decision making to account for non-human as well as human actors. The first stage is to identify propositions and entities to be taken into account, which can be human, non-human and hybrid actors. The second is to identify appropriate witnesses and stakeholders who can speak on behalf of the different actors and represent different interests. Scientists and engineers may have a particular role to play in identifying technologies or non-human nature as entities of interest and then speak on their behalf. The third stage is to arrange the different propositions and interests in hierarchical order, which accounts for human and non-human interests. The final stage is to stabilise the outcomes of the hierarchy of propositions and interests through new and existing public institutions. This process allows for human and non-human interests, and facts and values, to be considered and debated openly alongside one another, acknowledging that many environmental and technical issues may not be easily divided along conventional fact-value, nature-culture lines.

4.4   HYBRID ENGINEERING

Actor-network theory presents a wider view of the role of engineers in sustainable development than either ecological modernisation or radical ecology. Engineering is not merely a technical profession, but it involves negotiating relationships to form networks between human and non-human actors. The technologies and systems that engineers design, operate and maintain are not merely instruments for society and politics but shape social life and behaviour and stabilise values.

Actor-network theory provides a way of analysing existing systems in order to understand why patterns of unsustainable consumption remain so difficult to shift. Furthermore, it suggests that engineers have an important role to play in designing sustainability into new systems. Ecological efficiency may help reduce environmental impacts, but engineers can go further in designing for new social arrangements that support sustainability. Such design processes require participation by relevant human and non-human actors in order to succeed and to support democratic decision making rather than reverting to potentially autocratic models of 'social engineering.'

Engineering is a hybrid, socio-technical profession, and engineers have a central role to play in building actor-networks that cross over conventional divisions between society and technology, and nature and culture. Actor-network theory demonstrates that engineers must be aware of the risks of simplifying the interests of other actors when building relationships. Actor-networks can unravel if actors' interests are not adequately represented or addressed. Engineers therefore need to be more deeply aware of the needs and values of users as well as the performance of the technical components of their systems. Importantly, engineers need to be conscious that not all users or other humans in the network share their needs and values, and develop new methods of working that identify and account for different understandings of the purpose of new technologies and systems.

The recognition of engineering as a hybrid activity is consistent with recent representations of the profession and definitions of professional engineering competencies. The Institution of Civil Engineers believes that its members are "at the heart of society, delivering sustainable development through knowledge, skills and professional expertise". Successful engineering has always balanced economic and managerial objectives in delivering technologies and systems. Whilst engineering methods bring engineering science and technical best practice to solve problems, engineers are also involved in defining the scope of the problems and implementing solutions within society.

CHAPTER 5

Engineering Consumption

The ecological crisis is driven by unceasing growth in consumption of resources, due to increasing population and per capita consumption. Sustainable development will require reduction in per capita consumption of resources by the wealthiest people in order to accommodate a projected global population of 9 billion people, to account for a rise in consumption of the poorest people in the world and to conserve local resources and ecosystems. Achieving reductions in consumption and a high quality of life is perhaps the greatest challenge of sustainable development.

Consumption is both conspicuous and inconspicuous. Conspicuous consumption refers to purchases made by consumers in a deliberate way, which helps to form an identity, express an aspiration or reflect a set of values. Purchasing a car, clothing, books, music, leisure equipment, and even food can be examples of conspicuous consumption. Inconspicuous consumption refers to the resources we consume without necessarily noticing. Water, electricity and gas make up a considerable proportion of our daily resources consumption, yet are barely noticed (Shove, 2003). Engineers design, produce and distribute consumer goods which constitute conspicuous consumption. They are also largely responsible for the infrastructure which delivers water, energy and transport services, and removes waste from our homes and cities. The form of this infrastructure is central to how consumers use resources, but until recently, the role of infrastructure in shaping consumption was largely overlooked.

Infrastructure models based on 'predict and provide' approach to engineering forecast demand from consumers, based on projections of population and per capita consumption, and design infrastructure to meet it. Most modern infrastructure systems have been designed on the assumption of unlimited resources availability. Transport systems designed around the private automobile assume a limitless supply of affordable fuel. Water systems designed to deliver continuous high pressure supply assume endlessly renewable local water resources. Power systems based on continuous supply assume cheap, readily accessible fuel and continuously expanding generating capacity.

In recent decades, engineers' ability to provide resources at forecasted rates of growth has become constrained by limits on resources availability and infrastructure capacity. Power generating capacity has been overstretched in parts of the US. Gas prices have risen dramatically in parts of Europe and oil prices are vulnerable to political instabilities as remaining resources are concentrated in particular geographical regions. Water use was severely restricted in major Australian cities after almost a decade of drought at the beginning of this century.

Resource shortages shift attention from supply side models of infrastructure provision to demand management. This usually involves efforts to change consumer behaviour to reduce resource consumption and can also involve improving the efficiency of household technologies. Demand for infrastructure services such as energy, transport and water can be very difficult to change because consumption is so embedded in everyday life.

Infrastructure and consumption have co-evolved in everyday modern life (Shove, 2003, van Vliet et al., 2005). The provision of infrastructures services such as water and electricity allowed for vast improvements in basic living conditions and public health. However, the continuous provision of electricity and water to households also allowed the development of new technologies and practices which consume resources far above what is required for basic needs.

Consumption of resources, based on continuous supply from largely invisible infrastructure, has become embedded in everyday rituals and household practices such as taking a hot shower, flushing the toilet, watching TV or doing the laundry. These practices and the appliances they utilise have evolved along with the provision of continuous infrastructure services and have become part of 'normal life' for most modern citizens in the developed world and middle class and wealthy people in developing countries. Resources consumption in this context is barely noticeable amongst the everyday concerns of running a household and managing the stresses of modern life. As van Vliet et al. (2005) point out, the only interaction that most people have with the utility provider of infrastructure services is their bill, which arrives infrequently and is disconnected from the everyday practices that consume resources.

Engineers have been responsible for the provision of infrastructure and the resources that it delivers. The development of new uses for energy, transport and water by individuals, communities and society has not been of interest or concern to engineers. How households use resources provided by infrastructure systems has been considered to be a private matter, beyond the scope of professional engineering.

Alternative models of infrastructure provision involve different scales of resource extraction, generation and distribution, and greater attention to demand management. As demand management and smaller scale systems of provision become important, the details of consumption are of greater significance to engineers. Whereas the household was once considered a black box of consumption by engineers, they are increasingly being required to understand and intervene inside households in order to reduce consumption and develop new systems of production of energy and water, and new methods of waste management.

Sustainable engineering facilitates reduced resource consumption not simply through conventional ideas of ecological efficiency but also through shaping everyday routines and practices. Engineering understanding of the co-evolution of technology, infrastructure and everyday resources consumption can contribute to reconfiguring these relationships in ways that enable sustainability. The delivery of a 'service and flow economy' as part of the Natural Capitalism agenda discussed in Chapter 2 also requires new understandings of the importance of everyday life and new roles for infrastructure and technology.

5.1   MODERN INFRASTRUCTURE PROVISION

Electricity, water, sewerage, transport, communications and waste infrastructure are defining features of modern cities. Infrastructure has historically been provided by both the public and private sectors, in some cases changing between the two over time. For instance, in the UK, piped water was originally supplied by private companies, then in the 1890s water supply became the responsibility of government owned authorities, until the 1980s when water infrastructure was again fully privatised. Similarly, electricity, transport, gas and telephone infrastructure shifted between public and private ownership over the twentieth century in many countries. The public and private sectors share responsibility for some infrastructure provision, with public-private partnerships a common arrangement for building and operating roads and railway networks. In some instances, the public sector employs private contractors to provide infrastructure services, whilst maintaining ownership of the infrastructure assets themselves.

For most of the twentieth century, provision of universal access to basic infrastructure was a goal of municipal and national governments (Marvin and Graham, 2001). Access to water and sewerage were seen to be essential to good public health, and access to transport systems was essential to economic growth. As such, the state had a major role to play in the provision and management of infrastructure, which was seen to be a public good. Public investment in infrastructure was thought to be essential to the economic and social development. Infrastructure networks spread across cities and countries, linking citizens to the benefits of modern technology and resources consumption. Such investment depended on governments having ready access to finance to fund construction, and ongoing revenue to fund operation and maintenance.

Since the 1980s, the private sector has been increasingly involved in infrastructure provision. This has coincided with wider changes in public policy associated with privatisation and market liberalisation. The private sector is thought to now have better access than the public sector to financial markets to raise the capital necessary to invest in infrastructure construction and renewal, as well as being able to operate utilities more efficiently. Operating infrastructure as profit making enterprises shifts the emphasis away from the modern ideal of providing universal access to services which are public goods. Good regulation of utilities helps to maintain access to essential services for the poor. However, in rapidly growing cities that do not benefit from the legacy of well designed and constructed infrastructure, access to electricity, water, waste and transport services is found to be divided according to wealth (Marvin and Graham, 2001).

In many cities around the world it is not uncommon to have well connected, middle class suburbs with world class access to water, electricity, transport and communication services, side-by-side with slum settlements that have no formal provision. The challenge of providing basic services to the urban poor is made more complex by the often unplanned, illegal status of slum settlements and the difficulty of obtaining finance to build or extend infrastructure systems to low income areas. As a result, the poorest city residents often pay the highest prices for energy and water, resorting to private vendors who charge much higher prices than the centralised utility.

Modern infrastructure is largely based on centralised production or treatment, and networks for distribution. Water is abstracted from the environment, often outside city limits, stored in large reservoirs, treated in centrally managed treatment works, and distributed through extensive pipe networks to households, before being transferred to a network of sewers that transport wastewater back to centralised sewage treatment works before being released into the environment. Similarly, electricity is generated in power stations burning fuel outside the city, and distributed through networks of substations and power lines to individual homes and buildings for use. The systems of production and distribution are managed centrally by utility companies or authorities. Users have no interaction with these systems other than to consume the resources provided and to pay their bills. This allows for high levels of safety and security of supply, and it leaves users free to engage in modern life without concern for how the systems provide for their basic needs.

The design, construction and operation of large infrastructure systems have been some of the most important contributions of engineers to the development of modern society. Provision of clean water, sewers, electricity, gas and waste disposal to homes has freed modern citizens from concern about how to meet these basic needs and allowed them to engage in education, paid work, the arts, commerce, leisure, public debate, politics and other activities of modern life. However, such infrastructure has also facilitated consumption of resources far beyond the basic needs of most people and beyond the capacity of ecosystems to sustain.

5.2   INFRASTRUCTURE, CULTURE AND CONSUMPTION

Early environmental policies focussed on reducing pollution and improving efficiency of production systems. Since the 1990s, increasing attention has been paid to the role of consumption and consumer behaviour in achieving sustainability. Sustainable consumption usually focuses on the attitudes, behaviours and decisions of individual consumers, and the ecological efficiency of consumer technologies.

Focussing on the role of the individual consumer of resources has three key shortcomings: it underestimates the importance of shared cultural norms and values in driving patterns of consumption and overestimates the importance of individual knowledge and attitudes; it fails to recognise that most people use infrastructure for the services it provides in their daily life, not the resource it delivers (people use lighting, heating and refrigeration, not electricity); and it does not adequately account for the importance of infrastructure systems in shaping consumption. Infrastructures have allowed new technologies and practices to emerge which have changed everyday life and increased consumption of resources. Many of these changes have occurred relatively quickly, but without deliberate intention or awareness of their environmental consequences.

Elizabeth Shove in her book *Comfort, cleanliness and convenience* (Shove, 2003) points out the limitations of efforts to change individual consumption based on trying to change consumer attitudes to the environment and resources. Although consumption of energy, water and other infrastructure resources has significant environmental impacts, the use of these resources is so deeply embedded in everyday practices which are supported by widely held cultural norms that merely changing people's attitudes or providing more information is unlikely to change patterns of consumption. Consumption of such resources is inconspicuous, and secondary to concerns about social norms and everyday life.

Cultures of consumption have co-evolved with the provision of resources by modern infrastructure systems, changing social norms and technology. Shove (2003) presents the example of how the provision of continuous water supply to households enabled the development of the automatic washing machine to explain the processes of co-evolution and the ratcheting up of demand. As washing machines became more affordable and widespread, doing laundry became more convenient than before. Having a washing machine in your kitchen or bathroom is significantly more convenient than carrying clothes to and from a neighbourhood laundry or using older technology twin-tub or wringer washing machine. The added convenience of washing clothes using a household washing machine means that people are more likely to wash essentially clean but worn clothes that they previously would have considered re-wearing, airing or spot-cleaning. This contributes to raised social expectations that fresh clothes are worn each day or for different social occasions during the same day. The volume of clothes laundered on a regular basis and the volume of water used in households has consequently increased. Providing information about the environmental impacts of water consumption may have some influence on how often people use their washing machines, but it is unlikely to change individual and social expectations about the necessity for freshly laundered clothes. In this way, the convenience of laundering has led to increased expectations for cleanliness, which go far beyond requirements for public health and hygiene, and bear little relation to local water scarcity or the impacts of pollution on aquatic ecosystems. Similarly, the installation of showers in private bathrooms in the UK has shifted hygiene practices away from the weekly bath and daily sponge bath, towards daily showering, which increases water consumption and reflects changing social and cultural norms. Improving the water efficiency of showers or washing machines will reduce the volume of water consumed per use, but this is overwhelmed by increasing frequency of use of these appliances as cultures and expectations have changed. Furthermore, these expectations and norms are unlikely to shift simply by providing more information to individuals about the environmental consequences of their actions.

Information provided to consumers as part of demand management campaigns is often in conflict with the messages 'baked-in' to infrastructure systems. As infrastructures systems were designed with endless capacity to increase resource provision to meet demand, this assumption is evident in how they function. If a resident does not turn off the tap water, it will keep flowing down the drain. If they don't turn off the switch, the electric current keeps flowing to their appliances. Demand management campaigns that are based on informing users of the scarcity of resources are in contradiction to the message that the infrastructure systems themselves provide to consumers.

The scale of infrastructure provision is also in sharp contrast with the everyday experiences of users as they consume water, energy or other services. Zoe Sofoulis (2005) has coined the term 'Big Water' to refer to the scale and vision of water infrastructure provision and management in Sydney, Australia. Big Water consists of engineering systems and visions based on 'nation-building' dam construction projects, vast treatment works and distribution systems. This is in stark contrast to the 'Everyday Water' experience of household water use in Sydney. Water use is central to some of our most private and intimate activities – going to the loo, showering, bathing children or tending a vegetable garden (Allon and Sofoulis, 2006). Messages that 'Big Water' resources are running out and users need to change their behaviour to conserve water rarely account for the details of everyday water use. Furthermore, household fittings and appliances designed in line with the Big Water promise of endless supply are not easily adapted to support water saving efforts by conscientious householders. In studies by Sofoulis and others, water users have shown tremendous willingness to change their behaviours to conserve water, but the technologies and infrastructure of water provision are rarely available to support their efforts. Such changes in behaviour remain outside cultural norms without supportive technologies and infrastructures.

5.3   ALTERNATIVE INFRASTRUCTURE MODELS

Infrastructure in modern societies is facing several challenges. Infrastructure constructed in the nineteenth and early twentieth century is now reaching the end of its working life, requiring significant investment in maintenance, upgrading and replacement. Millions of people living in rapidly growing urban areas in developing countries do not have access to basic infrastructure services. The environmental imperative requires infrastructure providers to reduce pollution and resource consumption, and in particular to reduce carbon emissions. Climate change impacts will require infrastructure to adapt to improve resilience to extreme weather events. Demand management is increasingly important in infrastructure provision.

Discussions about alternative infrastructure models tend to contrast small scale, local systems with big, centralised provision. Proponents of decentralised systems of provision for water, energy, sanitation and waste services highlight resources losses in distribution as the basis for greater efficiency of local provision. Drawing on the work of E.F. Schumacher (1974), they also claim that decentralised systems are inherently more ecologically sound and socially beneficial. The debates between local and central infrastructure systems tend to overlook the possibility for intermediate scales of technology, networking of local provision, interconnections between large and small scale systems, and other opportunities to develop more sustainable modes of provision. Understanding the relationship between infrastructure, technology and cultures of consumption provides further opportunities for engineers to develop new ways of delivering the benefits of clean water, sanitation, energy and other infrastructure, whilst facilitating lifestyles and cultures that are not dependent on high resource consumption.

Changes in infrastructure provision in recent decades have been dominated by the rising importance of the private sector and the role of markets. Analysis of infrastructure provision since the 1980s has shown a trend towards 'unbundling' of utilities in line with increasing privatisation and consumer choice of providers (Marvin and Graham, 2001). Unbundling refers to disaggregation of generation, transmission, billing and other elements of infrastructure systems. Consumers are able to choose between different providers of services, who often share distribution networks. They are also able to benefit from entry of wholesaler providers of services, who do not own the infrastructure themselves, but are able to take advantage of bulk purchases from generators and then act as retailers to consumers (van Vliet et al., 2005).

Sustainable infrastructure is likely to include elements of centralised and decentralised systems. Systems that are critical for public safety, such as the provision of clean drinking water, may be best provided from centralised systems that can be managed and regulated efficiently. Systems that deliver lifestyle benefits rather than addressing basic needs may be provided by autonomous systems, for instance local collection and use of water for non-potable uses such as watering gardens. Centralised and decentralised systems use different technologies, but they also require different structures of management, ownership and regulation. Infrastructure can be provided on household, neighbourhood, city and regional scale, including connections between different scales.

Autonomous systems have contributed to the unbundling of conventional infrastructure systems. Consumers are now able to be the generators of infrastructure services, and as autonomous systems connect to the grid, they can provide these services to larger infrastructure providers and connect to distant consumers. Van Vliet et al. have represented the changes infrastructure provision as the differentiation of resources, providers and consumers. Resources were once non-renewable and used only once, and now include renewable and recycled resources. Monopolist providers have been replaced by a range of providers in the markets for services to consumers. Consumers are now differentiated by their values and interests in infrastructures services, with some people shopping around for the best prices for their utilities, while others are now able to express a preference for renewably sourced utility services despite increased cost. Furthermore, as small scale technologies such as photovoltaic cells and rainwater harvesting are more prevalent in households, consumers are themselves becoming providers of infrastructure services. In instances such as renewable energy production where individual households fitted with solar cells are able to feed in their excess electricity to the grid, households are providing resources to other providers of infrastructure services to consumers. Van Vliet et al. characterise these kinds of arrangements as 'co-provision' of infrastructure services, as the division between providers and consumers starts to blur.

Changing the scale of provision is also accompanied by changing the scales of regulation and management. Van Vliet et al. point out that the scale of management does not necessarily need to match the scale of technology. It is possible for small scale technologies to be installed in individual households and communities, yet managed by a large scale provider. Likewise, it is possible for community scale organisations to act as wholesalers in the networked provision of services purchased from large scale, central infrastructure resources. In mapping networks of institutions, users and technologies, relationships between social and technical actors can occur on a variety of scales.

5.4   ENGINEERING SUSTAINABLE INFRASTRUCTURE

Provision of infrastructure services involves appropriate financing, regulation, management and other institutional arrangements, as well as technology. Engineering expertise is often involved in all of these aspects. Engineers act as advisors to regulators and financiers, and infrastructure managers often have professional engineering qualifications. However, the relationships that shape consumption within households are often overlooked by engineers, planners, managers and regulators. Recent work in the sociology and geography of infrastructure and consumption is valuable in devising new models of provision, and they should be considered by engineers in technical design and operations, as well as in more managerial and advisory work.

Engineers have been central to the design management and operation of conventional infrastructure systems. Planners and governments have traditionally forecast economic and urban growth, and outlined demands for infrastructure, which engineers have then delivered. As infrastructure systems have been unbundled and resources, providers and consumers have become much more differentiated, the role of engineers is much less clear. Engineering techniques that have developed for large scale centralised systems are unable to adequately address the needs of consumer-providers in co-provision networks. Engineers have also struggled to devise appropriate demand management strategies or to account for potential resource savings from demand management measures. These require engineers to have a greater understanding of users and their relationships with technologies and infrastructure systems. Sociologists, demographers, psychologists and other social scientists have important knowledge to bring to bear on these problems, but understanding how this impacts on technology and infrastructure design and management requires knowledge of technical systems, which is the usual domain of engineers.

Designing new systems for sustainable systems of infrastructure requires work at multiple scales, and an understanding of how they intersect. Engineers are working to develop renewable energy sources and water recycling systems that will provide new resources for conventional centralised utility systems. Engineers are also working to improve the efficiency of distribution systems and reduce losses. Improving the resource efficiency of household technologies is important, but needs to take into account users needs and practices, and be conscious of the rebound effect where improvements in efficiency are undermined by increased use.

Managing resources and infrastructure across different scales of production and consumption requires new engineering systems. Engineers have an important role to play in analysing and optimising systems at different scales, to deliver the most sustainable system possible rather than falling prey to dichotomous debates between the inherent efficiencies and benefits of centralised versus decentralised technologies. Engineers have traditionally undertaken this analysis on purely technical terms, but if such analysis is to support sustainability, then the relationships with social and cultural elements of the system must be considered. Systems shape cultures and behaviours, and so comparison of different systems based on the same social assumptions may not be valid.

CHAPTER 6

Sustainable Urban Water Systems

The provision of clean water and sewerage services to modern cities is one of the greatest achievements of the engineering profession. Adapting those systems to the challenges of climate change and growing populations, and providing water and sewerage systems to the millions of people currently living without adequate access, are two of the greatest challenges facing the profession in the twenty-first century. Water is one of the most essential and basic resources needed to support life, and using water is associated with some of our most private activities. Consumption of water is determined by biological, personal, social, technological and environmental factors, and production of clean water and disposal of wastewater have significant environmental impacts which need to be managed. Treating and distributing water and wastewater around cities requires vast amounts of energy, and as a rule of thumb, it is generally the case that the largest customer of an electricity utility in a city is the water utility. Massive concrete dams across some of the world's largest rivers to provide water for agriculture and cities, and hydro-electric power for industrial development, are icons of modern engineering. For these reasons, the challenges of urban water systems provide a useful case study of the relationships between engineering, society and sustainability.

Global freshwater resources are highly constrained and under increasing stress. While agriculture is the greatest consumer of water resources, sustainable provision of water to cities is a growing challenge, as discussed in Chapter 1. Water is heavy and incompressible, and therefore difficult to transport over long distances. Cities are limited by locally available water resources. While agriculture dominates water use globally, locally urban water use can have significant impacts on resources and aquatic environments. Cities that have developed water systems based historical rainfall patterns are now adapting to climate change, which is predicted to change the distribution of rainfall throughout the year as well as annual average rainfall in many cities and their water catchments.

Provision of water and sanitation services to the urban poor in rapidly growing cities in the developing world is an issue of increasing concern. Urban water poverty is usually the result of lack of access to infrastructure or affordable sources of safe water, not due to absolute scarcity of water (Allen and Bell, 2011). Within the same urban area, middle class and wealthier residents may have continuous supply of clean water in their households and consume hundreds of litres of water per capita per day, while the urban poor exist on very marginal, often illegal, sources of supply. With more than half the world's population now living in cities, and urban populations forecast to continue to grow, devising affordable and sustainable models of infrastructure provision is essential.

The challenge of devising sustainable urban water systems is complex. Sustainable access will mean an increase the levels of consumption by the poorest people who do not currently have access to sufficient volumes of safe water to meet their basic needs for health and hygiene. It will require a reduction in demand for uses that are far beyond what is required to meet basic needs, and changing cultural and lifestyle expectations that require high levels of water consumption. Sustainable water systems will also require the development of new water resources that do not harm local ecosystems and are energy efficient. Sustainable urban water systems need to be designed according to local hydrological, economic, social and environmental conditions. Whilst specific technologies may be transferrable between different cities, the structure and operations of urban water systems will change according to local conditions.

Big dams

Massive concrete dams across major rivers have been some of the most iconic achievements of the civil engineering profession and have been highly contentious in many places. Dams provide water resources, flood protection and hydro-electricity but also result in the destruction of ecosystems in the area dammed and disrupt hydrology and ecology downstream. Dam construction also often displaces people who live in the flooded area, creating local opposition and social justice concerns as the most negative impacts of dam construction are usually borne by people who are unlikely to benefit from it. In 2000, the World Commission on Dams produced guidelines for good practice in dam projects *Dams and Development,* which outlined five core values – equity, sustainability, efficiency, participatory decision making and accountability (WCD, 2000).

6.1   MODERN WATER INFRASTRUCTURE

Modern urban water infrastructure abstracts water from the environment, often from outside the city, stores it in dams and reservoirs, treats it to a high standard to remove bacteria and pollutants, and distributes it through a vast network of pipes to individual buildings, businesses and homes. City residents then use the water, often for merely a few seconds as it runs from the tap, toilet or shower head, before it flows into a network of sewers, to be treated at a sewage treatment works to remove bacteria and pollutants, before being discharged back into the environment, often beyond the city limits. This vast infrastructure is essential to good public health in cities. However, the linear flow of water from the environment, through the city where it is used for just a few moments, and back into the environment at a different point is increasingly unsustainable.

The provision of universal access to safe drinking water and sewerage in cities was originally justified on the grounds of public health. During the nineteenth century, environmental conditions in rapidly industrialising cities of Europe and North America declined considerably (Melosi, 2000). Traditional systems for removing human waste and accessing clean water were overwhelmed by dramatic population growth. Local water sources became polluted by human and industrial waste, increasing the transmission of diarrheal disease. Cholera epidemics had devastating impacts in cities such as London, where it claimed 37,000 lives between 1831 and 1866 (Halliday, 1999).

The accepted theory of disease transmission for most of the nineteenth century was the theory of miasma, which proposed that disease was caused by bad smells. The germ theory of disease attributed to Louis Pasteur and Robert Koch was not established until the 1870s. The emerging science of bacteria and its relation to disease was not fully established until the end of the nineteenth century. Public debate about the poor state of public health and engineering design of water and sanitation infrastructure was mostly based on the miasma theory of disease. The work of Dr. John Snow in London showed that there was a link between contaminated water and cholera cases, but the dominant view was that disease was caused by breathing in contaminated air, rather than drinking contaminated water. Famous British sanitary reformer Edwin Chadwick wrote that 'all smell is disease', as the foundation for promoting water and sanitation infrastructure that would prevent disease by washing away bad smells and the rotting matter from which the stench arose.

The main objective in providing households with continuous flowing water was therefore to wash away decaying matter that created bad smells, or miasma, which caused disease. This required households to be connected to urban drainage systems, which had been built for removing surface water but which were being converted into sewers to wash away foul smelling waste from houses and factories. The invention and adoption of the water closet, or flushing toilet, during the middle of the nineteenth century typifies the growing use of water as a medium for washing away smelly waste from households in order to prevent disease. Sewerage systems were consequently required to ensure the smelly, contaminated water was transferred away from the city.

Alternative systems for dealing with miasma and rotting waste were proposed that did not require continuous flows of water. The 'earth closet' competed with the 'water closet' as a technology to safely remove organic waste but allowed it to be more easily utilised as manure (Geismar, 1993). Systems for removing excrement based on vacuum or compressed air pressure were developed in the UK and the Netherlands, with the objective of improving public health by maintaining circular flows of nutrients (Geels, 2006). Such propositions were not widely implemented, and water based sanitation systems with extensive sewer networks came to characterise modern cities.

Despite being based on a faulty scientific theory, the basic design of urban water infrastructure systems was successful in improving public health and ending cholera outbreaks in most European and North American cities. Even before the bacterial theory of disease was widely accepted, water treatment technologies such as sand filtration, which removes bacteria and other particles, were implemented (Melosi, 2000). The construction of sewers removed the cause of bacterial contamination of local water sources, reduced exposure and stopped the spread of disease. The principles of continuous clean flowing water and household connections to sewers were established as the basis for public health engineering.

Household connections to continuous, high pressure water systems, and drainage to take away wastewater, allowed for the development of uses of water that were unforeseen by the public health campaigners of the nineteenth century. New bathing practices and standards of personal hygiene developed, which initially contributed to improved health but increasingly became about personal comfort and standards of cleanliness that are socially rather than biologically determined. Elizabeth Shove's (2003) research shows the importance many people now attach to a deep bath or shower as a readily available source of relaxation in an otherwise busy modern life, as discussed in Chapter 5. Laundering clothes changed over the century as washing technologies and products developed, and mass produced clothing allowed for many more changes of clothes and more frequent laundering, with concomitant changes in social norms and expectations about appropriate dress and cleanliness.

Increasing demand from new uses for water and growing urban population was met by expanding water infrastructure. For most of the twentieth century the 'predict and provide' model of infrastructure provision dominated, with engineers expanding supply systems to meet demand. In this model of water provision, the household is a black box, water consumption and water using practices are a private matter. The role of engineers is to predict aggregate demand based on demographic, economic and other variables, then to build systems that are capable meeting it.

This approach is still firmly based on a commitment to maintain good public health. Whilst most water supplied is used for purposes that are not directly related to health, a loss of supply would place public health at serious risk. Maintaining critical supplies of safe drinking water to households is an essential function of water utilities and cannot be compromised.

6.2   HYDROLOGICAL LIMITS

In recent decades, the predict-and-provide model of centralised provision of water and sewerage services has faced significant challenges. Local water resources are fully exploited in many catchments that supply cities, yet populations continue to grow in most urban areas. Environmental protection legislation has placed greater emphasis on maintaining the quality and quantity of water in streams, lakes and wetlands, constraining the exploitation of water resources to sustainable levels. Water and sewerage networks that were built in the nineteenth and twentieth centuries require increased investment in maintenance, upgrades and replacement. Land-use conflicts over dams and reservoir construction constrain options for building new infrastructure.

Water infrastructure managers now pursue a twin-track approach of exploring options for expanding supply and reducing per capita demand. Water resources cannot be indefinitely expanded, and reducing per capita demand relieves pressure on existing supplies due to population growth.

Reduced per capita demand can help expand the number of people supplied by current resources and postpone or cancel the need for investment in new supplies.

In regions where capacity for abstracting water from the environment is highly constrained and demand is likely to exceed supply in the short and medium term, desalination is being implemented as a new source of water. Desalination has been a major source of water for several decades in the Middle East and in some remote locations, but only recently has it been used more widely. The major limitation of desalination is the energy required to treat the saline water to produce fresh water. Early desalination technology was based on distillation, and was highly inefficient. More recently, membrane filtration technologies, particular reverse osmosis, have been developed for desalination. Reverse osmosis involves applying high pressure to saline or brackish water to force fresh water through a membrane, leaving a more concentrated brine solution behind. The energy requirements to achieve the pressures required to force the water through the membrane are very high.

Continued improvements in membrane technology have resulted in improved energy efficiency for desalination, but compared to conventional sources of water, this remains a highly energy intensive and expensive water supply option. In some recent developments such as in London and the Western Australian city of Perth, desalination plants have been powered by renewable energy resources. This helps alleviate immediate concerns about increased carbon emissions due to desalination but highlights concerns about the sustainability of desalination as a significant component of water supply systems, given the broader challenge of reducing overall energy consumption and carbon emissions from cities. Using renewable energy to supply new energy intensive processes such as desalination is better than using fossil fuels. However, sustainability requires constraining growth in energy demand and using renewable to replace existing fossil fuel generation, not simple to slow growth in fossil fuel use. In this context, desalination is not a sustainable solution for water supply, even if it is supplied by renewable energy.

Indirect Potable Reuse (IPR) is another alternative supply option which has been explored and implemented by some water utilities. This involves treating wastewater to a very high standard, then introducing it back into conventional water sources (river, aquifer, reservoir) before treatment through the normal drinking water systems. The technology at the heart of IPR is reverse osmosis or micro-filtrations, which are similar to desalination. The pressures required to separate the clean water from pollutants in treated wastewater are less than for desalination, but the energy requirements for these treatment technologies remain significantly higher than for conventional water treatment processes. The public acceptability of IPR is of concern in many cities, and proposals for IPR have been rejected by the public in Australia and the USA. IPR remains a less energy intensive option for new supplies than desalination, but it has been subject to more intense public controversy (Bell and Aitken, 2008).

6.3   MANAGING DEMAND

As conventional water resources are constrained and new sources of water are expensive, controversial and energy intensive, attention has turned to reducing per capita demand for water. Demand management can take many forms including legal restrictions on water use, education campaigns and improving the efficiency of water using appliances. Demand management activities are most common during periods of water stress, but they are also important to achieve longer term, stable reductions in per capita water use.

During periods of extreme water stress, such as prolonged drought, water utilities may restrict certain uses of water. This usually involves limits on outdoor water use, from only allowing outdoor water use on particular days of the week, to banning the use of sprinklers and hosepipes. Most city water authorities have a staged approach to water restrictions, with restrictions becoming more severe as water resource conditions worsen. Engineers are involved in assessing water resources and deciding when to implement restrictions on use. They also consider the impacts of the different actions to save water in determining the types of restrictions to implement at various levels of water stress when planning ahead for how to manage drought conditions.

Education campaigns are also common elements of water demand management strategies. Providing information to the public, customers and through schools is thought to improve willingness to conserve water. Education campaigns often involve alerting people to the problem of water stress, the costs of wasting water, and ideas for simple behaviour change. Campaigns have focussed on choosing drought tolerant plants for home gardens, taking shorter showers and turning the tap off when brushing teeth. There is little evidence to support the theory that simply providing people with more information leads them to make lasting changes to their behaviour. Water using behaviours are often part of an unconscious daily routine and are associated with cultural norms about cleanliness that are hard to change. Education campaigns which support water restrictions during drought periods can be highly effective in the short term, but water using behaviour and per capita consumption usually reverts to higher levels when the drought is over.

Improving the efficiency of household appliances and fittings is important in water demand management. Education campaigns may be accompanied by provision of small water saving devices for users to install in their homes. These include cistern displacement devices, which can be placed in toilet cisterns to reduce the volume of water stored in the cistern and used each flush. Low flow shower heads are also a common device provided to households for free or at a subsidised price to reduce the flow rate of water from showers. Water utilities or governments provide subsidies for people to install rainwater tanks or rainwater butts to collect water from roofs for outdoor use. More extensive demand management campaigns involve replacing existing fittings with more water efficient devices, such as the toilet replacement programme in New York which installed more than 1 million water efficient toilets in three years during in the 1990s (USEPA, 2002). Households can be encouraged or subsidised to replace existing washing machines and dishwashers with more water efficient models. Providing water efficiency information by labelling such devices is an important to allow consumers to take account of water efficiency in their purchasing, but this will only be one element of their purchasing decision.

Water efficiency can also be promoted through building codes or standards for new constructions. Regulation and specification of water efficient devices in new homes and buildings can be achieved through voluntary as well as compulsory standards. Plumbing standards that have previously been based entirely on public health concerns are now being adapted to incorporate water efficiency measures. Specifications can apply to individual devices or fittings, or overall calculations of building water consumption, which allows designers flexibility in meeting overall standards for consumption. The UK Code for Sustainable Homes is an example of a building code which allows flexibility in how designers choose to meet set standards for water efficiency in new homes. In the Code, the overall per capita consumption of the house design is calculated based on the assumed use and performance of fittings, and the designer can choose between a range of water efficiency measures to meet standards required for different levels of rating in the Code scheme (Communities and Local Government, 2011).

Water metering can help reduce water demand. In most of the UK and in other countries, households are charged a flat rate for their water service irrespective of how much water they use. Their water use is not measured, and they have no economic incentive to reduce their consumption. Water metering allows consumers to measure their use and pay for the volume they use. With appropriate billing information, they can also see how their use changes over time or compares with average use in their local area. In order for water metering to have a significant impact on water demand, it needs to be accompanied by appropriate water pricing. Rising block tariffs, for instance, have increasing charges per unit of water as the level of consumption increases. Customers are charged a low unit price for the first 'essential block' of consumption, then an increasingly higher unit costs for 'non-essential' use, with unit prices rising in steps as consumption increases. Essential uses of water are low cost, luxury uses of water are charged at a premium. This still allows for basic needs to be met at a relatively low cost, but it provides penalties for increasingly profligate use. Water charges might also be varied seasonally to allow utilities to charge more during times of water shortage to further encourage households to reduce their use.

Setting charges for water is a very contentious social and economic issue. Larger families may be less wealthy, though charged high prices for high total water use, even though larger households have lower per capita consumption. Single person households use the highest volume of water per capita, but may remain below the threshold for rising block charges. Elderly or frail people may have particular needs for high water usage, such as kidney dialysis or taking long baths or showers to relieve pain, and again may be least able to afford high water charges. Although water charging may appear to have a straight forward economically rational impact on water consumption, unless pricing structures are thoughtfully devised, they can have unfair social impacts.

Attention to individual behaviour, pricing and water efficient technologies in managing demand are useful starting points but fail to address the importance of relationships between technology, infrastructure, culture and consumption. Zoe Sofoulis (2005) has pointed out the inconsistencies between an engineering culture of 'Big Water' and the intimacy of daily experiences of water. Elizabeth Shove's (2003) research has demonstrated the importance of social and cultural expectations which shape everyday water using practices, and have co-evolved with technologies and infrastructures. Achieving significant, long term reductions in per capita demand for water requires redesigning water systems to account for the connections between culture, technology, infrastructure

> **Free basic water in South Africa**
>
> The national constitution of South Africa includes the right of all citizens to sufficient water. In 2001, the national government implemented a free basic water policy to provide free supply of at least 25 litres per person per day within 200m of their home. Households that consume more than 25 litres per person per day pay for their consumption, usually according to a rising block tariff. In cities such as Durban, the cost of free basic water to the peri-urban poor is covered by cross subsidies from higher consuming water users in wealthier suburbs (Muller, 2008).

and water using practices. This requires reconfiguring infrastructure and household systems to not only conserve water but also to shift expectations and practices that lead to high water consumption.

6.4   APPROPRIATE INFRASTRUCTURE

Although based on the same essential objectives of providing continuous supply of water and removal of wastewater, the actual design and operation of urban water infrastructure is highly specific to the context of the city. Water infrastructure systems are designed for different rainfall patterns and water resource availability. They also reflect the history of the city, urban form, economy, local politics and technical and administrative capacity.

Sustainable water systems will require reconsidering the appropriate scale of provision of water, based on local conditions and the purpose that water is used for. In the UK, around one third of drinking water is used for flushing toilets. Abstracting water from the environment, treating it to a high standard, pumping it across the city, to flush it down the toilet may not be an appropriate use of a limited resource. Sustainable water systems need to consider public health, water conservation, energy conservation and environmental impacts. Designing sustainable systems should look for opportunities to eliminate water use, improve water efficiency, use water of an appropriate quality and re-use water, while maintaining basic water and sanitation services to ensure good public health. Involving the public and users in decision making and design is also important to achieve sustainable and acceptable outcomes. Optimising energy, water and environmental impacts to achieve sustainability requires strong quantitative assessment, and it may require tradeoffs between different sustainability objectives.

Engineering expertise is essential to devising sustainable urban water systems at different scales. However, this expertise must be applied across scales and in ways that take account of social and cultural factors. Whilst the 'big engineering' contribution to urban water and sanitation infrastructure has achieved great advances in public health and quality of life, it has also resulted in significant

Scripting hot water using behaviour

Domestic hot water systems are an example of how the design of technologies reenforce particular assumptions and behaviours. Hot water consumption is an important issue for reducing energy as well as water use. In the UK, hot water consumption is the second highest energy consuming activity in households after space heating (Energy Saving Trust, 2009). Reducing hot water use is therefore important to meet targets of reduced carbon emissions as well as conserving water.

There are two basic options for domestic water heating – continuous boiler or storage heater. The continuous boiler heats water as it flows to the hot water tap. The storage heater heats a set volume of hot water over a longer period. Under normal operation, users of a continuous hot water system should never run out of hot water. Users of a storage heater will run out of hot water once the stored volume has been used. Continuous boilers are generally more energy efficient than storage boilers, but allowed users to stand under a hot shower for as long as they choose. Storage boilers, on the other hand, implicitly limit the use of hot water within a household at any given time to the volume of the boiler.

Storage boilers have also been used by energy network managers as potential devices for storing energy. Low night time tariffs encouraged customers of a UK energy company to time the main heating of their water in the middle of the night. This tariff was introduced to reduce the peak demand for energy early in the morning, but it created an unexpected new peak demand in the middle of the night as customers took advantage of the lower energy prices (van Vliet et al., 2005).

ecological destruction and contributed to the evolution of unsustainable patterns of consumption. 'Big engineering' will always be important in water infrastructure provision, but further expertise is required at different scales, as well as good understanding of how the different scales of provision interact, and the co-evolution of technology, infrastructure and cultures of consumption.

6.4.1   ELIMINATE

Eliminating the use of water wherever possible is an important starting point for achieving sustainability. In some situations, such as the provision of sanitation in developing countries eliminating water use is essential. With almost 50% of the world's population living without access to improved sanitation, it is practically impossible to connect every household to a water-borne sanitation system. There is simply not enough water for every house to have a flushing toilet, nor enough money to construct an extensive sewerage and sewage treatment system. Waterless sanitation is the only option in many situations.

Changing daily practices can be as simple as using a broom to sweep down outdoor areas, rather than hosing down with high pressure water. Temporary water restrictions during drought events eliminate non-essential outdoor water uses using legal controls. Designs for new appliances can consider how to achieve the benefit that users seek from water using practices, without consuming water itself. For instance, appliances that 'refresh' clothes by airing or heating can give users the experience of fresh clothes without washing them. Replacing lawns in gardens in dry climates with more appropriate drought tolerant plants and 'xeri-scape' garden designs can eliminate the need for watering. The Southern Nevada Water Authority ran a successful campaign in the city of Las Vegas which involved paying customers to remove turf from their gardens, thus providing the incentive and funding for households to plant drought-tolerant plants, eliminating or reducing the need for watering.

6.4.2   REDUCE

Reducing water use in existing activities is a key challenge for demand management, involving improved efficiency of devices, changing behaviour of users and a better understanding of the relationships between technology and behaviour. Improving the water efficiency of appliances must not be at the expense of reduced performance, or there is a risk that overall water consumption will remain high. For instance, low flush toilets that do not clear the toilet bowl are likely to be flushed twice instead of once; washing machines that do not rinse clothes properly may result in rinse cycles being run again; and people may stand under a low flow shower for longer to rinse shampoo from their hair. Design and installation of water efficient devices should also consider potential rebound effects. If people are aware that their appliances are water efficient, they may use them more often, negating improvements in efficiency. For instance, if users know that their toilet has a low flush volume, they may flush it unnecessarily to dispose of household waste; they may be less reluctant to wash relatively clean clothes in a water efficient machine; and they may stand under their low flow shower for longer.

6.4.3   REUSE

Reusing water can be done on a variety of scales. It can be as simple as washing up in a bowl and pouring the dishwater on a pot plant, or having two sets of pipes in a housing development for potable and reused water. When reusing wastewater, it is important to distinguish between black water and grey water. Black water refers to sewage, which is contaminated with faecal matter, including flows from toilets. Grey water is relatively clean wastewater from other household activities such as showering and washing clothes.

Grey water reuse is increasingly applied at the household scale. Simple grey water reuse systems involve directing rinse water from washing machines onto gardens. More complex systems include storage, treatment and plumbing to use grey water for toilet flushing as well as outdoors. Maintaining grey water systems to ensure safe and efficient operation can be a problem at household scale, requiring active monitoring and management by homeowners. Alternatively, these activities could be contracted out to a service provider, to install, inspect and maintain the system as a fee for service. Grey water systems operating over several households in medium and high density developments, such as blocks of flats, could be better managed by building managers, allowing safety and security of grey water supply without relying on individual homeowners to become experts in water management.

Grey water systems with limited potable backup can encourage users to reduce water. If grey water is the only source of water for gardening, without any backup from the central water system, then households are more likely to manage their use of non-potable water and plant selection to ensure long term availability for outdoor use. In this way, grey water systems can encourage behaviour and cultural change as users are more directly aware of the sources and limits to water supply.

Reusing black water requires a much higher level of treatment and control than for grey water. Consequently, these systems do not usually operate at household scale and are most often found at neighbourhood scale. Sewage can be treated using conventional technologies at centralised sewage treatment works or using newer decentralised technologies such as membrane bioreactors. Water reuse can again be as simple as directing treated wastewater on to playing fields, or re-distributing it back through neighbourhoods to provide a supply of non-potable water in a separate set of pipes to the conventional potable supply. Houses in such neighbourhoods are plumbed with two sets of pipes which are usually different colours. One set of pipes is for potable or high contact uses such as drinking, cooking and showering. The second set of pipes delivers reclaimed water for uses such as toilet flushing, gardening and laundry.

The rebound effect is also evident in water reuse systems, with householders less likely to implement changes to reduce their water consumption if they believe they are using reclaimed water. This can lead to an overall increase in per capita water use. If reclaimed water is unavailable for technical reasons or if all available reclaimed water has been consumed, and non-potable supply is automatically replaced by potable water, then this can undermine benefits of installing the water reuse system (Livingston et al., 2005).

6.4.4   SUBSTITUTE

Rainwater harvesting can be another source of non-potable water at building and neighbourhood scale. Again this can be as simple as collecting rainwater from gutters in a small barrel for use in water plants outdoors, or it can involve installation of tanks, pumps and dual plumbing systems for non-potable supply within the house, building or neighbourhood. Rainwater harvesting can have the additional benefit of attenuating storm water flows, reducing flows into local drainage systems during rainfall events which can alleviate local flood risks and reduce pollution of local waterways.

Rainwater harvesting is preferred over grey water as it is of higher quality; however, the supply is less certain as it is dependent of rainfall patterns rather than the usage of water within the building. As a result, rainwater storage tanks are usually significantly larger than the storage required for grey water, but rainwater requires much less treatment to ensure safety for non-potable uses.

6.4.5   RECYCLING

Recycling water at larger urban scales takes the form of direct or indirect potable reuse. All the water distributed to houses is still of potable quality, with treated wastewater an additional supply option for the centralised water infrastructure system. Potable reuse has the advantages of minimal changes to the overall water systems of the city, but treatment is energy intensive and the technology has not been acceptable to the public in some cities. The fact that potable reuse is essentially invisible to water users means that it is unlikely to facilitate behaviour or cultural change.

6.5   URBAN DRAINAGE

Provision of clean drinking water and removal of sewage from cities are two key elements of water management and engineering. A third important aspect is the drainage of surface water from rainfall. Good drainage is essential for health, safety and effective functioning of a city. Strategies for designing and managing drainage systems in cities have undergone significant change in recent years.

The essential aim of conventional drainage systems is to remove any water that falls on the city as quickly as possible and to ensure drains are big enough to be able to cope with a 1 in 5 or 10 year rainfall event. Surface water can either flow into the sewerage system and be treated as wastewater, in a combined sewer system, or can flow through a separate drainage network and be released into the environment with minimal treatment. Whether a city has a separate or combined sewer system depends on its historical and economic context with combined sewers more prevalent in older cities that were built before a modern sewerage system was installed. Newer cities tend to have separate drainage and sewerage systems. Combined sewers have the disadvantage that during large rainfall events, the sewerage system cannot cope with all the wastewater and storm water and dilute sewage overflows into local waterways, causing significant pollution. In cities where land is increasingly paved or built over, runoff to drains during storm events has increased, as less water is able to infiltrate into soils. As urban surfaces become less permeable, the likelihood of drainage systems failing causing flooding or pollution in increased for the same sized rainfall event in a more permeable catchment (Butler and Davis, 2004).

Sustainable urban drainage and water sensitive urban design are new approaches to designing and managing storm water. Whereas conventional drainage engineering aims to remove all the water from the local area as quickly as possible, sustainable systems enhance opportunities for infiltration, storage, evaporation and use of storm water. Techniques include removing concrete paved surfaces, using ponds and swales to store and infiltrate water, rainwater harvesting, green roofs, infiltration devices and permeable paving. These techniques can also provide opportunities to increase vegetation and biodiversity in urban areas, with benefits for wildlife and reduced urban heat island effect. They also indicate increasing consideration of water issues beyond engineering, in urban planning and design.

6.6   WATER POVERTY

Lack of access to water in cities is usually due to economic, social and political conditions rather than hydrological conditions. In most cities, there is enough water available to meet people's basic needs. The fact that millions of people do not have access is largely the result of failure of governments, financiers, utility managers and engineers, rather than lack of water resources (Allen and Bell, 2011).

The urban poor use many different strategies for accessing water in the absence of functioning infrastructure. Some may access water from a communal standpipe, carrying water home and storing it in a variety of containers. The standpipe may be connected to the central water infrastructure or local ground water resources. Others buy water from vendors, who deliver water to slum areas using tankers or smaller vehicles. Water kiosks provide water in cities such as Dar es Salaam, and bottled water may be used by those who can afford it. Illegal connections to central water supply systems are common in many cities. Illegal access to water can be an individual action or part of an organised criminal activity. Some households share water connections, with neighbours who have access providing water to those who don't. Rainwater harvesting can also provide a local source of water for households.

The quality of water from these different sources is highly variable, and households may treat water before drinking it or using for cooking. This usually involves boiling, which can consume a large proportion of the household's fuel budget and be very expensive. Other techniques include household filtration and chemical disinfection, using local resources and traditional techniques or proprietary products.

Many of these different strategies and techniques are very expensive compared with the price of water paid by those who have a connection to the central utility water network. Paying water charges to utilities is a highly contentious issue for many poor communities, particular as water is often believed to be a shared resource and water provision a basic public service. However, the cost of water from alternative sources can be much more expensive. The cost of poor health due to insecure water supply is a further burden on already vulnerable households.

Water and sanitation provision in urban areas can be analysed in terms of needs driven and policy driven approaches (Allen et al., 2006). Needs driven approaches directly address the immediate needs of people for water, and include informal and sometimes illegal methods. Policy driven provision on the other hand focuses on more formalised methods for water provision, organised according to different approaches to governance, management, ownership and regulation of water infrastructure.

Allen et al. (2006) argue for including needs based approaches in formal planning and management of water infrastructure, rather than focussing exclusively on the provision of centralised infrastructure services. Intermediate provision, starting with the needs of the urban poor, can lead to faster improvements in quality and quantity of access than focussing exclusively on the construction of large scale systems. On the other hand, intermediate provision can divert attention to longer term infrastructure solutions. With governments under pressure to meet targets for water provision, communities who are deemed to have 'access', though substandard, may be excluded from future infrastructure programmes.

6.7   ENGINEERING SUSTAINABLE WATER SYSTEMS

Through the provision of clean drinking water and sanitation, the engineering profession can claim to have saved more lives than the medical profession. This is an achievement of which engineers should rightly be proud. However, the professions must also carry some of the shame that this infrastructure fails to reach nearly a billion people across the world and that it has led to the destruction of mighty rivers and catastrophic degradation of aquatic ecosystems. Engineers are at the forefront of finding sustainable solutions to these persistent problems.

Sustainable water systems will not be achieved using the same model of engineering that has created the problems we now face. Sustainable engineering will not be focussed entirely on large, centralised water systems delivering potable water across the city as if it was an unlimited resource. Sustainable engineering will consider new models of infrastructure provision that facilitate behavioural and cultural change to reduce demand for water in wealthy households and meet the basic needs of the poor. These systems are likely to operate across scales, including household and neighbourhood as well as city wide infrastructure. Engineering expertise must be extended to incorporate better knowledge of how technology and society interact, and brought to bear in designing and operating new systems. Minimising energy use and environmental impact, as well as maximising resilience to increasingly uncertain climatic conditions, make engineering sustainable water systems more complicated than conventional predict-and-provide models of infrastructure provision. Engineers working on sustainable water systems are involved in setting building codes for water efficiency, designing demand management campaigns and water efficient devices, installing and maintaining rainwater harvesting and grey water recycling systems, managing neighbourhood non-potable water schemes, and planning for water security into the future. All of these tasks involve knowledge of social, economic and political systems as well as traditional technical analysis. Engineering sustainable water systems is a socio-technical undertaking reflecting the hybridity of urban water systems.

CHAPTER 7

Engineering, Technology and Ethics

Sustainability presents a particular challenge to the engineering profession. Moving beyond ecological modernism and efficiency, to more fundamental changes in the relationships between humans, technology and nature requires a new understanding of engineering and the impacts of technology. Engineers are the professionals most associated with technological systems and engineering decisions made during design, construction and manufacture of new technologies have lasting implications for the use and lifecycle of technologies and infrastructure. Engineers and philosophers have reflected extensively on the ethical implications of engineering actions and decisions. Sustainability, like many engineering problems, is a moral as well as a technical challenge. Designing ecologically efficient systems can be a presented as merely a technical problem for engineers, but it is part of a wider ethical challenge that requires engineers along with the rest of society to question their values in relation to human exploitation of nature and the continued failure of industrial development to deliver basic services like water and sanitation to the world's poorest people.

The modern engineering profession has its origins in the military. Engineers were originally responsible for the construction of roads, bridges and the production of armoury in standing armies (Davies, 1998). As the industrial revolution began in the seventeenth century, engineers were also employed to build canals and new transport routes for increased trade, as well as to oversee the design and implementation of new industrial technologies in textile mills. By the beginning of the nineteenth century, engineers who had risen from various backgrounds, including the trades and crafts as well as the educated middle class, began to form professional associations and develop standards for professional education and training. The Institution of Civil Engineers was formed in London in 1828 as the world's first professional engineering institution. The designation 'civil' engineers distinguished the work of these new engineers from their military counterparts. Over the following century, new disciplines of engineering formed as technology developed and engineering knowledge became more specialised. Mechanical, chemical and electrical engineering emerged in the nineteenth century, with electronic, environmental, software and many other disciplines forming in the twentieth century.

7.1 ENGINEERING KNOWLEDGE

Vermaas et al. (2011) in their volume in this series analyse the role of engineers in designing discrete technologies and more complex socio-technical systems. Engineers are central to the development, design, manufacture, construction, operation, maintenance and disposal of both technology and socio-technical systems. They are also involved in policy making, financing, regulating and managing technology and socio-technical systems. Engineers bring particular specialist knowledge about technologies and infrastructure to these arenas, but, more importantly, engineers bring a particular way of knowing the world. Engineering knowledge and method are the defining elements of engineering as distinct from other professions and academic disciplines. They underpin the different disciplines of engineering, and they are what engineers have in common across domains as broad as nuclear reactors, computer chips, steel works, satellites, drug manufacturing and racing cars.

Engineering knowledge stems from individual and collective experience of solving physical problems, with a specific purpose in mind. Engineers use and develop tools, techniques, and methods for solving problems more efficiently and with greater certainty. These include mathematics and scientific theory, but engineering is not merely the practical application of maths and science. Indeed, scientific knowledge is often derived from engineering problem solving. Engineering knowledge is based on constantly extending the 'state of the art,' which refers to the best available knowledge and techniques for analysing physical problems and designing solutions. Engineering knowledge must be adaptable to specific conditions of new contexts in which problems occur.

Heuristics and 'rules of thumb' are an important element of engineering knowledge (Koen, 2003). Rules of the thumb are convenient simplifications that are agreed upon by the professional community in order to solve new problems as efficiently as possible. Whilst it is important for engineers to consider the specific context and conditions of every new project or problem, it is not necessary to thoroughly analyse each situation from first principles. Nor is it necessary to be able to derive detailed explanation for why a particular tool or rule is effective, if collective experience attests to its effectiveness. Engineers are most interested in what works and how it works, without requiring a detailed scientific explanation of the physical principles underpinning the phenomena they are exploiting or seeking to control (Vermaas et al., 2011). Detailed scientific understanding may help extend the state of the art and provide insights into the nature of problems at hand, but it is not a necessary precondition for good engineering per se.

Standardisation and codification are important for stabilising and extending engineering knowledge. Based on scientific investigation and practical experience, engineering codes help stabilise rules of thumb and provide minimum standards for safe performance. Agreed through professional peer-review processes, engineering standards address methods as well as outcomes.

The relationship between engineering and the physical sciences is well established, and engineers are expected to have a good knowledge of basic scientific principles as well as a familiarity with the scientific method. Engineers are generally less familiar with the social sciences and social science methods. As a result engineers' knowledge of society is often based on personal experience, the popular media or stereotypes. Engineers use rules of thumb relating to social phenomena similar to those relating to physical phenomena. However, social rules of thumb are less likely to be subject to critical evaluation or the processes of standardisation that are important for safety and the ongoing development of the state of the art.

Engineers rarely work alone and are usually associated with large organisations, either being directly employed by large industrial corporations, the government, military and consulting firms, or subcontracting to them. Working collectively helps engineers share experience and knowledge, and provides checks on individual work to ensure safety. Team work also allows for specialisation within engineering projects, so that individuals bring detailed understanding of 'state of the art' knowledge across different domains.

Engineers' association with large projects and organisations is also one of the core characteristics of the profession and a significant challenge to the capacity of the profession to respond to the challenges of sustainable development (Davies, 1998, Riley, 2008). As technologies and systems for sustainability operate across different scales and address the needs and interests of users, the environment and the poor, it may become necessary for the profession to develop new models of professional formation, expertise and service. Incorporating local knowledge into engineering decisions, providing engineering expertise to support community initiatives for sustainability, and designing systems that engage with everyday life of users to reduce resources consumption are significant challenges to the conventional model of engineers associated with large scale developments and organisations (Baillie, 2006).

Engineers' tendency to work in teams and as part of large organisations, and the importance of standardisation and codification of engineering work, means that their design work is rarely recognised by society in the same way that the work of product and architectural designer is acknowledged. Whilst historical figures are recognised within and beyond engineering, the engineering profession do not produce celebrity figures in the same way that architecture or other professions do. Engineering attention to safety, team work and fulfilling the needs of the client tends to mitigate against the rise of particular personalities. Engineering design, though highly creative and innovative, is often invisible to users of technologies and systems. Engineers work to create stable black-boxes. Engineering design most commonly only becomes visible when systems fail, and good engineering is often by its nature unnoticed by everyday users and the public.

7.2   TECHNOLOGY AND VALUES

Debates about technology and values often concern the extent to which technology determines human actions and social outcomes. Andrew Feenburg (1999) has identified two key dichotomies in ways of thinking about technology. The first is the dichotomy between technology as value neutral or value laden, and the second is between technology as human controlled or autonomous. Presented as a matrix, these lead to four distinct characterisations of the nature of technology (Table 7.1).

Most engineering approaches to technology can be characterised as instrumentalist. Engineers characteristically maintain that technology itself is value neutral, and that humans express values through how they use technology. Humans are in control of technology and technological development. Technological progress is humanly determined, and whether or not it leads to positive of negative outcomes for society or the environment depends on how technologies are used, not the

Table 7.1: Theories of Technology (Feenburg, 1999, p. 9)		
Technology is:	**Autonomous**	**Humanly Controlled**
Neutral (complete separation of means and ends)	Determinism (e.g., traditional Marxism)	Instrumentalism (liberal faith in progress)
Value-laden (means form a way of life that includes ends)	Substantivism (means and ends linked in systems)	Critical Theory (choice of alternative means-ends systems)

nature of the technologies themselves. A very crude simplification of the instrumentalist view of technology is the slogan 'guns don't kill people, people do.'

In addressing the moral and social as well as technical elements of sustainability, engineering can learn from more critical approaches to technology and socio-technical systems. As actor-network theory shows, technologies are both shaped by and determine social relationships and human behaviours. Technologies have particular assumptions about the world 'baked-in' to them, and as such perpetuate the worldviews that formed those assumptions. Centralised water infrastructure systems were based on assumptions of endless supplies of water from the environment, resulting in cultural practices that demonstrate a very low value of water and aquatic ecosystems that may be harmed as a result of over abstraction. Technologies such as automatic washing machines represent values of cleanliness and convenience, but they do not necessarily value water as a scarce resource.

Critical theory presents technology as value laden but humanly controlled. This means that although technology embodies certain human values, humans control technology and technological development through design, regulation and operation (Feenburg, 1999). Throughout the modern industrial age, when environmental values were not considered important, technologies were developed that reflected values of economic growth and higher standards of living. As values change, it is possible for engineers and others to become engaged designing, building and operating technological systems that embody the values of sustainability. These values include preserving natural systems and resources for their own sake, as well as for future generations, and addressing the needs of the world's poor through appropriate development.

There are different views as to the role of values in design. Friedman and Kahn (2003) identify three different worldviews. An 'embodied' view of design holds that designers inscribe particular values and intentions into the technology. The 'exogenous' view is that societal forces shape how technology is used, consistent with an instrumentalist view of technology. The 'interactional' conception of design is that design supports some values and hinders others, but that the user has ultimate control of the technology.

The interactional view of design is broadly consistent with an actor-network approach which shows how networks of humans and non-humans are stabilised but can also unravel. Technologies script certain behaviours that reflect particular values, but these depend on meeting the interests of users (Vermaas et al., 2011). Users are free to re-script technologies with their own values. For example, continuous clean water has been provided by engineers to households in the interest of public health, but users in networks of appliance, plumbers and manufacturers have re-scripted water infrastructure as a source of pleasure and cleanliness.

7.3   ENGINEERING VALUES

The extent to which values influence engineering work depends on underlying assumptions about the nature of technology and design. An instrumentalist view of technology and an exogenous perspective on design absolve engineers from any direct responsibility for the uses of technology and social or ecological consequences. Even from more critical perspectives on technology and design, the engineer can still be seen as a servant of client values. Alternatively, as technical experts, engineers can be seen as gate keepers of technology, as central actors in technological society who allow the material expression and stabilisation of particular values and the suppression of others. Engineers can also be thought of as mediators between technology and society, facilitating the interactions between values and objects that shape technology.

Engineers as servants simply carry out the instructions of clients, and they have no control over, hence no responsibility for the outcomes of their work. Clients' demands may be constrained or shaped by wider social and political values, either directly through legislative controls, or indirectly through social norms and aspirations, but the practice of engineering is separate to these processes. Engineers' primary responsibility is to their client, who bears responsibility for the outcomes of technology within the wider context of social and political norms and regulations. Clients can be the corporate employers of engineers, governments, or private individuals. The engineer has no power to influence or change the rationale or outcomes of technology but merely implement the demands of more powerful clients.

Engineers as gatekeepers hold ultimate responsibility for the outcomes of their work. Heroic accounts of successful engineering projects such as the construction of London's sewers, the Hoover Dam, space exploration, and the computer revolution, put engineers centre stage. In these instances, engineers are ultimately responsible for the positive social outcomes of technological progress. On the other hand, engineers must also carry responsibility for failures of technology and technical systems. Engineers have been held responsible for spectacular technical failures such as the space shuttle Challenger disaster and the Bhopal chemical gas catastrophe. Although engineers may be acting in the interests of their clients, their primary responsibility is to the health and safety of the public, and as gatekeepers of technology, they ultimately determine the values that can be expressed through design, development and operation of technology.

Engineers as mediators are critical actors arranging networks of humans and non-humans into technologies and socio-technical systems. Engineers work to mediate the interests of different humans within the realm of physical possibilities. Engineers cannot design technologies that are physically impossible but are constantly expanding the realm of the possible through their intimate knowledge of the non-human actors and the state-of-the-art that they work with. Aligning the non-human actors with human interests is a primary task for engineers, and as actors themselves in the network, they have significant capacity for negotiating these relationships. However, engineers' agency is constrained by what is physically possible, in terms of the limits of non-human actors, and what is socially, politically and economically achievable, in terms of maintaining the interests of primary human actors.

Engineering codes of conduct typically recognise engineers as servants, gatekeepers and mediators simultaneously. The gatekeeper role of engineers is evident in their primary ethical responsibility to ensure public safety. The servant role is demonstrated through their duty to act in the best interests of the client. The mediator role is increasingly important as codes of conduct have been expanded to include a duty to work to achieve sustainable development.

7.4   MEDIATING SUSTAINABILITY

Engineers as mediators have the potential to reconfigure relationships between humans and non-humans to ensure the long term viability of natural resources and systems for future generations, and to acknowledge the intrinsic value of non-human nature. Engineers as mediators can recast dualistic relationships of domination of culture over nature, bringing the interests of non-human natural actors to the networks of socio-technical systems. However, this requires significant reform of the role of engineers in society and the economy and new models of engineering practice. New techniques for engineering design have been developed that account for the role of values in technology development and allow for greater involvement of the public in technical decision making. These include value-sensitive design and constructive technology assessment.

Value-sensitive design methods have been specifically developed in software engineering but have wider applicability to sustainability engineering. Friedman and Kahn (2003) outline three elements of value-sensitive design methodology: conceptual, empirical and technical investigations. Conceptual investigation involves designers asking questions about what stakeholders are directly or indirectly affected by the design, what values are at stake, and what are the likely tradeoffs between values. Empirical investigations involve engineers drawing on social science methodologies to move beyond conceptual questions to gather data regarding the actual interests and values of those implicated in the technology under development. Methods for gathering social data include surveys, interviews and focus groups. Empirical data about stakeholder values and interests can be considered alongside technical data in the design of the technology. Technical investigations focus on how existing technologies support or suppress the expression of particular values, and, in turn, how new technologies can be consciously developed with particular values in mind. The most significant contribution of value sensitive design is to acknowledge that technologies embody particular values and to bring this to the consciousness of designers. This allows for greater ethical deliberation by engineers as they consider their role in configuring particular arrangements between technology, society and nature. Engineers may not always have the power to promote the values of sustainability within a design project, but these methods allow them to be more cognizant of their limits as well as the opportunities to achieve change.

Constructive Technology Assessment has developed in the Netherlands, Denmark and elsewhere in Europe (Rip et al., 1995). Technology Assessment originated in the 1970s as European governments became concerned about the impacts of new technologies on society and the environment. Methods of evaluating new technologies as input to regulation were developed, including citizen panels, socio-technical experiments and expert analysis. Citizen panels take randomly chosen or self-selected members of the public, provide them with detailed information about new technologies and access to technical experts, and ask them to provide an assessment of the likely impacts and acceptability of new specific new technologies. Socio-technical experiments involve small scale trials of technology in use. Assessment of new technologies provided government with guidance in developing new technology policies and regulations but was based on essentially stabilised technologies, late in the lifecycle of development.

Constructive Technology Assessment extends the principles and techniques of Technology Assessment upstream, to influence the design of new technologies, rather than simply assessing the impacts of more-or-less finalised products. This allows for a wider range of social and policy goals to be incorporated into new technologies, and it helps to dismantle the traditional divide between technological development and social impacts. Constructive Technology Assessment allows engineers to obtain feedback on their work early in the stage of development, so that they are able to respond to concerns about impacts and incorporate aspirations that they may not have considered independently. Constructive Technology Assessment explicitly acknowledges the hybrid, socio-technical nature of most technology development projects, and it provides opportunities to consider and direct technological development to maximise positive outcomes and minimise harm.

These design tools break through the separation of technology and society which has characterised the development of modern industrial society. Conventionally, engineers are isolated from society, designing and managing technologies according to the requirements of clients, corporations and governments. This model of technological progress allowed for the proliferation of new technologies and delivered significant advances in the material standards of living of millions of people. However, it also created lifestyles that required the consumption of resources at unsustainable rates, industrial systems that have polluted the environment and destroyed ecosystems, and failed to deliver even the most basic benefits of development to the world's poorest people.

7.5   SUSTAINABLE ENGINEERING

Sustainable development acknowledges the need to reconsider the dominant model of modern industrial development to address social, ecological and environmental considerations simultaneously. Sustainable development is a hybrid concept, which is neither technical nor social but bridges across these distinctions. Engineering contributions to sustainable development to date have been restricted to providing technological solutions to environmental problems and have struggled to address social dimensions, and the relationships between technology, environment and society.

Philosophers of technology, sociologists and anthropologists have argued that technology is value laden. The major infrastructure systems that underpin modern societies, including transport, energy, water and waste, embody assumptions that natural resources are endless, that supply can continually be extended to meet demand, that the environment can absorb pollutants, and that natural systems have no value other than for human use. New technologies and infrastructure systems as a starting point need to question these assumptions and values. As engineering codes of conduct now explicitly refer to engineers' ethical obligations to society, the environment and the goals of sustainable development, new technologies and systems should be developed that embody alternate values.

Engineers have significant influence in the design and management of technological systems, but they work in economic and political contexts in which economic growth, profit and ever increasing material standards of living predominate. Engineers alone will not be able to deliver sustainable technology. The ecological crisis will not be solved simply by engineers producing more efficient, less polluting technologies. However, as mediators between technology, society and nature, engineers are uniquely positioned to present alternatives to dominant models of development and to facilitate the emergence of more sustainable patterns of consumption.

CHAPTER 8

Conclusion

Humans are now changing the environment on Earth at an unprecedented rate. Carbon emissions are increasing global temperatures and changing the climate. More than half the world's original forest cover has been lost or modified. One third of the world's species are vulnerable or threatened with extinction. A quarter of the world's arable land is severely degraded. Conventional oil production has peaked in many regions and is likely to peak globally in the next few decades.

The human population on Earth is now 7 billion and predicted to stabilise at 9 billion by 2050 compared with 1 billion in 1800. More than 2 billion people do not have access to improved sanitation, and more than 700 million do not have access to safe drinking water. By 2050 more than half the world's population will live in water scarce regions.

The environmental crisis is a result of increasing population and rates of consumption of resources. Populations are growing fastest in developing countries, and developed countries have the highest per capita rates of consumption of energy, water, food and other resources. The processes of industrial development which are responsible for the overexploitation of natural resources and pollution of the environment have delivered tremendous benefits to billions of people, but these have failed to reach the poorest billions who still live in degraded environments without access to basic infrastructure and services.

Sustainable development has been proposed as the route to meet the needs of current generations while protecting the environment and conserving resources so that future generations can meet their needs. Sustainable development aims to meet the needs of the poor and protect the environment. In strong formulations, it involves protecting and restoring the environment for its own sake, not simply for the future use of humans. Weak versions are merely the incorporation of pollution control and more efficient technologies in existing models of industrial development.

Engineering has been central to modern industrial development that has delivered tremendous benefits but has also led to unprecedented ecological destruction. Engineering is essential to sustainable development. However, sustainable development presents significant challenges to the profession. Sustainable engineering has mostly focussed on improving the environmental performance of existing industries, and as such can be associated with weak models of sustainable development and the dominant policy agenda of ecological modernisation. Stronger versions of sustainable development will require attention to the human and social elements of sustainable development, which conventional methods and models of engineering have largely ignored.

Sustainable development is a cultural as well as a technical challenge. More immediately, sustainability will require significant changes in everyday life and cultures in high consuming societies. At a more fundamental level, the exploitation and destruction of the environment indicates a deep ethical failure of modern western culture. Environmental philosophers have used historical and cultural scholarship to explore the reasons for human domination of nature and propose radical alternatives to current patterns of development, politics, human settlement and community.

The rapid development of science and technology which have been so central to modernity has been made possible by the separation of nature and culture. Nature has been studied by science and controlled by technology for the benefit and progress of human culture. Knowledge about the nature and technology has been separated from knowledge of the economy, law, society and culture. This allowed scientists and engineers to rapidly develop detailed specialist knowledge and solve specific problems, but it also meant that any issues that fall between specialist disciplines have been left unresolved and have grown in complexity. The human impact on the environment and sustainable development are examples of such issues that fall into the gap between nature and culture.

Actor-network theory is a method of analysis which does not break the world into the primary categories of nature and culture. It analyses science, technology, society, culture and everything that crosses between them on the same terms. Actor-network theory describes the world in terms of networks of human and non-human actors. It describes the processes of building and unravelling of actor-networks. Human action can be prescribed by non-human actors, and human intentions can be undermined by failure to adequately account for the reality of non-humans. Actor-network theory provides a useful way of analysing problems of sustainability as they occur, rather than from predetermined disciplinary perspectives. It shows the interaction between social and technical elements of problems and the construction of solution and demonstrates the fallacy of the idea of a 'technical fix' to most complex problems. Engineers are very important actors in building networks of humans and non-humans that are capable of sustainability and are not based on the conventional exploitation and domination of nature.

The provision of infrastructure is a defining feature of modern societies and developed economies. Transport, energy, water and waste infrastructure underpin modern life and are generally taken for granted, until they are disrupted or fail. Infrastructure provision is one of the core tasks of the engineering profession. Engineers are responsible for designing, constructing, operating and maintaining infrastructure systems, and they have significant input into planning, managing, financing and regulating infrastructure.

Infrastructure systems developed in the nineteenth and twentieth centuries have followed the 'predict and provide' model, assuming the capacity of the system and the resources it requires can be indefinitely expanded to meet rising demand. These assumptions were 'baked in' to the system and sent users the message of limitless energy, water and land resources, and it relieved users of concerns about the impacts of their actions. Product designers, manufacturers and users developed new uses for energy and water in households, and everyday cultures and behaviours co-evolved with the adoption of new technologies and the provision of infrastructure services. This has resulted in ever increasing consumption of resources, the impacts of which remain invisible to users but have profound environmental consequences.

This book began by asking the question *how can engineers help build a sustainable society?* Achieving sustainability will require designing and redesigning infrastructures and technologies which enable the co-evolution of low consumption behaviours and cultures. This is an essential task for engineers, and requires them to reflect on the values and assumptions that are embodied in the technologies they design. It is likely that engineering expertise will be required at different scales than the conventional large scale infrastructure and organisations that engineering has relied upon in the past. New methods for design are developing that allow engineers to take fuller account of knowledge of society, including local knowledge of users. Value-sensitive design provides a framework for engineers to reflect on the stakeholders and values involved in their work, including identifying key tradeoffs between competing values. Constructive technology assessment considers the potential impacts of new technologies throughout the design process, enabling opportunities to address adverse outcomes or enhance positive impacts and aspirations from the earliest conceptual stages of a project.

Sustainability is widely acknowledged as a core responsibility of professional engineers. Codes of conduct for engineers now consistently include reference to sustainable development. Engineering organisations are clear that sustainability involves social as well as environmental considerations, and engineering contribution to sustainable development is not limited to developing more efficient, less polluting technologies. However, engineers also have responsibilities to act in the interests of their clients, and their primary ethical responsibility remains the safety of the public. Sustainable development should not conflict with safety responsibilities. However, as part of an economic and political system that has the primary goal of maintaining economic growth, which remains closely tied to increasing consumption of resources, engineers' obligations to sustainable development and their clients may increasingly come into conflict.

Engineers are problem solvers. They are essentially pragmatic, not idealistic or dogmatic. Engineers throughout history have traded off their ambitions for technical efficiency and elegance with the immediate demands of clients or the constraints of the economic or environmental context in which they operate. Engineers are not mere servants of their clients. Engineers are powerful actors in the construction of socio-technical networks, but they remain mediators rather than commanders. Engineers have the capacity to work with clients to make incremental changes towards sustainable development and will continue to improve the efficiency and reduce the environmental impacts of all projects they take on. They can work with governments and policy makers to enhance regulation and stimulation of sustainable development within their fields of expertise. Engineers can also promote new technologies and systems that work across a range of different scales to address needs that are not currently met by large scale systems. Engineering expertise can be broadened to include more knowledge of society and culture and their relationships with technologies. Even where sustainable development is not achieved, engineers can be clear about the tradeoffs that have been made and the values that have been stabilised in technologies and systems.

Sustainability requires changes in how humans interact with natural systems and with each other. This cannot be achieved simply by continuing existing patterns of development in the hope that new technologies will solve environmental pollution and overcome resource shortages. Since the industrial revolution, most cultures and regions have witnessed dramatic social, technical and economic change. The historical trajectory of industrial and technological development cannot continue. Engineers have a central role in determining the direction of change in response to growing ecological and social crises. Ensuring a positive contribution requires that engineers recognise the social as well as environmental implications of their work and help to establish the conditions for the co-evolution of sustainable socio-technical systems.

Bibliography

3M Corporation (2011) 3P- Pollution Prevention Pays, online `http://solutions.3m.com/wps/portal/3M/en_US/3M-Sustainability/Global/Environment/3P0`, accessed 24 July 2011. Cited on page(s)

Allen, A. and Bell, S. (2011) Glass half empty? Urban water poverty halfway through the Decade of Water for Life *International Journal of Urban Sustainable Development* 3(1) 1–7. DOI: 10.1080/19463138.2011.583530 Cited on page(s) 61, 73

Allen, A., Dávila, J., and Hofmann, P. (2006) The peri-urban water poor: citizens or consumers? *Environment and Urbanization* 18(2) 333–335. DOI: 10.1177/0956247806069608 Cited on page(s) 73

Allenby, B.R. (1999) *Industrial ecology: policy framework and implementation.* Upper Saddle River, New Jersey: Prentice Hall. Cited on page(s) 2

Allon, F. and Sofoulis, Z. (2006) Everyday Water: Cultures in Transition *Australian Geographer* 37(1) 45–55. DOI: 10.1080/00049180500511962 Cited on page(s) 58

Aumônier, S., Collins, M., and Garrett, P. (2008) *An updated life-cycle assessment study for disposable and reusable nappies* Environment Agency Science Report – SC010018/SR2, on-line `http://randd.defra.gov.uk/Document.aspx?Document=WR0705_7589_FRP.pdf`, accessed 28 July 2011. Cited on page(s)

Bailey, J. (1997) Environmental Impact Assessment and Management: An Under-explored Relationship *Environmental Management* 21 317–327. DOI: 10.1007/s002679900032 Cited on page(s)

Baillie, C. (2009) *Engineering and society working towards social justice: Part I, Engineering and Society* San Rafael: Morgan and Claypool. DOI: 10.2200/S00136ED1V01Y200905ETS008 Cited on page(s) 45

Baillie, C. (2006) *Engineers within a local and global society* San Rafael: Morgan and Claypool. DOI: 10.2200/S00059ED1V01Y200609ETS002 Cited on page(s) 43, 77

Baker, S. (2006) *Sustainable Development* London: Routledge. Cited on page(s) 18, 43

Barry, J. (2005) Ecological Modernisation, in Dryzek, J.S. and Schlosberg (eds) *Debating the Earth* 2nd edition, Oxford: Oxford University Press, 303–321. Cited on page(s)

Bell, S. and Aitken, V. (2008) The socio-technology of Indirect Potable Water Reuse *Water Science and Technology: Water Supply* 8(4) 441–448. DOI: 10.2166/ws.2008.104 Cited on page(s) 65

Bell, S., Chilvers, A., and Hillier, J. (2011) The socio-technology of engineering sustainability *Proceedings of the Institution of Civil Engineers: Engineering Sustainability* available online 27 June 2011. Cited on page(s) 2

Benyus, J.M. (2002) *Biomimicry* New York: Perennial. Cited on page(s) 2

Bijker, W. (1997) *Of bicycles, bakelites and bulbs* Cambridge: MIT Press. Cited on page(s) 46

Bookchin, M. (1995) *The Ecology of Freedom* Oakland: AK Press. Cited on page(s) 38

Bookchin, M. (1988) Social Ecology Versus Deep Ecology *Socialist Review* 18: 9–29. Cited on page(s) 38

Butler, D. and Davies, M. (2004) *Urban drainage* 2nd edition, London: Spon. Cited on page(s) 72

Callon, M. (1986a) The Sociology of an Actor Network: the Case of the Electric Vehicle, in Callon, M., Law, J., and Rip, A. (eds.) *Mapping the Dynamics of Science and Technology* London: The Macmillan Press Ltd, 19–34. Cited on page(s) 46, 47

Callon, M. (1986b) Some elements of a sociology of translation: domestication of the scallops and the fishermen of St Brieuc Bay, in Law, J. (ed.) *Power, Action and Belief. A new sociology of knowledge?* London: Routledge and Kegan Paul, 196–233. Cited on page(s) 46

Callon, M., Law, J., and Rip, A. (1986) *Mapping the Dynamics of Science and Technology* Hampshire: Macmillan Press. Cited on page(s) 46

Carson, R. (2002) *The Silent Spring* 40th anniversary edition, Boston: Houghton Mifflin Company. Cited on page(s) 14

Chambers, R., Pacey, A., and Thrupp, L.A. (1989) *Farmer First* Rugby: Practical Action. Cited on page(s) 18

Clark, J. (1998) A social ecology, in Zimmerman, M., Baird Callicott, J., Sessions, G., Warren, K., and Clark, J. (eds) *Environmental Philosophy* 2nd Edition, Upper Saddle River: Prentice Hall, 416–440. Cited on page(s) 39

Communities and Local Government (2011) Code for Sustainable Homes, on-line `http://www.communities.gov.uk/planningandbuilding/sustainability/codesustainablehomes/`, accessed 25 July 2011. Cited on page(s) 67

Costanza, R., D'Arge, R., De Groot, R., Farber, A., Grasso, M., Hannon, B., Limburg, K., Naeem, S., O'Neill, R.V., Paruelo, J., Raskin, R.G., Sutoon, P., and Van Den Belt, M. (1997) The value of the world's ecosystem services and natural capital *Nature* 387(6630) 253–260. DOI: 10.1038/387253a0 Cited on page(s) 7

Costello, A., Abbas, M., Allen, A., Ball, S., Bell, S., Bellamy, R., Friel, S., Groce, N., Johnson, A., Kett, M., Lee, M., Levy, C., Maslin, M., McCoy, D., McGuire, B., Montgomery, H., Napier, D., Pagel, C., Patel, J., Puppim de Oliveira, J. A., Redclift, N., Rees, H., Rogger, D., Scott, J., Stephenson, J., Twigg, J., Wolff, J., and Patterson, C. (2009) Managing the health effects of climate change *The Lancet* 373(9676) 1693–1733. DOI: 10.1016/S0140-6736(09)60935-1 Cited on page(s) 13

Canadian Society for Civil Engineering, Institution of Civil Engineers, American Society of Civil Engineers (CSCE, ICE, ASCE). (2006) *Protocol for engineering: A sustainable future for the planet,* online `http://email.asce.org/ewri/December2006.html#Sustainability`, accessed 30 July 2011. Cited on page(s) 2

Daly, H. (1977) *Steady-State Economics: The Economics of Biophysical Equilibrium and Moral Growth* San Francisco: W. H. Freeman and Co. Cited on page(s) 17

Davies, M. (1998) *Thinking like an engineer* Oxford: Oxford University Press. Cited on page(s) 75, 77

Dobson, A. (2000) *Green Political Thought* 3rd edition, London: Routledge Cited on page(s) 42

Ebisemiju, F.S. (1993) Environmental Impact Assessment: Making it Work in Developing Countries *Journal of Environmental Management* 38 247–273. DOI: 10.1006/jema.1993.1044 Cited on page(s)

Ehrlich, P. (1968) *The Population Bomb* San Franscisco: Sierra Club - Ballantine Books. Cited on page(s) 15

Energy Saving Trust and Environment Agency (2009) *Quantifying the Energy and Carbon Effects of Water Saving* London: Energy Saving Trust. Cited on page(s) 69

Engineering Council UK (2009) *Guidance on Sustainability for the Engineering Profession* online `http://www.engc.org.uk/about-us/sustainability`, accessed 30 July 2011. Cited on page(s) 2

Food and Agriculture Organisation (FAO) (2010) *Aquastat* online `http://www.fao.org/nr/water/aquastat/main/index.stm`, accessed November 2010. Cited on page(s) 9

Feenburg, A. (1999) *Questioning Technology* London: Routledge. Cited on page(s) 77, 78

Foreman, D. (1993) *Ecodefense: a field guide to monkeywrenching* 3rd Edition, Chico CA: Abbzug Press. Cited on page(s) 38

Friedman, B. and Kahn, P.H. (2003) Human values, ethics and design, in Jacko J.A. and Sears A. (eds) *The human-computer interaction handbook* Mahwah NJ: Lawrence Erlbaum Associates, 1177–1201. Cited on page(s) 78, 80

Geels, F. (2006) The hygienic transition from cesspools to sewer systems (1840–1930): The dynamics of regime transformation *Research Policy* 35: 1069-1082. DOI: 10.1016/j.respol.2006.06.001 Cited on page(s) 63

Geismar, J. (1993) Where is Night Soil? Thoughts on an Urban Privy *Historical Archaeology* 27(2) 57–70. Cited on page(s) 63

Gleik, P. (2008) *The World's Water: 2008–09* Washington: *Island Press.* Cited on page(s) 9

Global Footprint Network (2010) *Footprint for nations* online http://www.footprintnetwork. org, accessed 22 July 2011. Cited on page(s) 14

Green Belt Movement (2006) What is the Green Belt Movement, online http:// greenbeltmovement.org/, accessed 30 July 2011. Cited on page(s) 11

Halliday, S. (1999) *The Great Stink of London* Gloucestershire: Sutton Publishing Ltd. Cited on page(s) 63

Hardin, G. (1995) The Tragedy of the Commons, in Conca K., Alberty M. and Dabelko G. (eds) *Green Planet Blues Boulder:* Westview Press, 38–45. Cited on page(s) 15

Hawken, P., Lovins, A.B., and Lovins, L.H. (2000) *Natural Capitalism* London: Earthscan. Cited on page(s) 2

Hay, P. (2002) *A Companion to Environmental Thought* Edinburgh: Edinburgh University Press. Cited on page(s) 38, 39

Huber, J. (2000) Towards Industrial Ecology: Sustainable Development as a Concept of Ecological Modernization *Journal of Environmental Policy and Planning* 2 269–285 DOI: 10.1080/714038561 Cited on page(s)

Huber, J. (2004) *New Technologies and Environmental Innovation* Cheltenham: Edward Elgar. Cited on page(s)

International Energy Agency (IEA) (2010) *World Energy Outlook 2010* Paris: International Energy Agency. Cited on page(s) 12

Intergovernmental Panel on Climate Change (IPCC) (2007) *Climate change 2007. The physical science basis,* Contribution of working group I to the fourth assessment report of the intergovernmental panel on climate change. Solomon S., Qin D., Manning M. *et al.* (eds) Cambridge: Cambridge University Press. Cited on page(s) 13

IUCN (2006) *RedList of Threatened Species,* online http://www.iucnredlist.org/, accessed 22 July 2011. Cited on page(s) 12

Johnson, P.M., Mayrand, K., and Paquin, M. (2006) *Governing Global Desertification* Aldershot: Ashgate Publishing. Cited on page(s) 11

Koen, B.V. (2003) *Discussion of the Method* Oxford: Oxford University Press. Cited on page(s) 76

Latour, B. (1993) *We Have Never Been Modern* Essex: Pearson Education Ltd. Cited on page(s) 4, 44, 47, 48

Latour, B. (1987) *Science in Action* Cambridge MA: Harvard University Press. Cited on page(s) 46

Latour, B. (1998) *The Pasteurization of France* Cambridge MA: Harvard University Press. Cited on page(s) 46

Latour, B. (1991) Technology is society made durable, in Law J. (ed.) *A Sociology of Monsters: Essays on Power, Technology and Domination* London: Routledge, 103–131. Cited on page(s) 49

Latour, B. (2004) *Politics of Nature* Cambridge MA: Harvard University Press. Cited on page(s) 50

Latour, B. and Woolgar S. (1979) *Laboratory Life* London: Sage. Cited on page(s) 46

Law, J. (1986) On the Methods of Long Distance Control: Vessels, Navigation, and the Portuguese Route to India, in John Law (ed.) *Power, Action and Belief: A New Sociology of Knowledge?* Henley: Routledge, 234–263. Cited on page(s) 46

Leopold, A. (1970) *A Sand County Almanac: with other essays on conservation from Round River.* New York: Ballantyne Books. Cited on page(s) 36

Livingston, D.J., Stenekes, N., Colebatch, H.K., Waite, T.D., and Ashbolt, N J. (2005) Governance of water assets: a reframing for sustainability *Water* August 19–23. Cited on page(s) 71

Markandya, A., Harou, P., Bellu, L.G., and Cistulli, V. (2002) *Environmental Economics for Sustainable Growth* Cheltenham: Edward Elgar. Cited on page(s)

Marvin, S. and Graham S. (2001) *Splintering Urbanism* London: Routledge. Cited on page(s) 18, 55, 58

Maslin, M. (2008) *Global warming, a very short introduction.* Oxford: Oxford University Press. Cited on page(s) 13

McDonough, W. and Braungart, M. (2002) *Cradle to Cradle* New York: North Point Press. Cited on page(s)

McHarg, W. (1992) *Design with Nature* 25th anniversary edition, New York: John Wiley and Sons. Cited on page(s)

Meadows, D., Randers, D., and Meadows, D. (2005) *Limits to Growth. The 30-Year Update*, London: Earthscan. Cited on page(s) 17

Meadows, D.H., Meadows, D.L., Randers, J., and Behrens, W.W. (1972) *The Limits to Growth* New York: Universe Books. Cited on page(s) 15

Melosi, M. (2000) *The Sanitary City* Baltimore: The Johns Hopkins University Press. Cited on page(s) 63, 64

Merchant, C. (1990) *The Death of Nature* 10th Anniversary Edition San Francisco: Harper. Cited on page(s) 40

Mol, A.P.J. (1995) *The refinement of production: Ecological modernization theory and the chemical industry* Utrecht: International Books. Cited on page(s)

Mol, A.P.J. and Sonnenfeld D.A. (2000) *Ecological Modernisation Around the World* London and Portland: Frank Cass. Cited on page(s)

Muller, M. (2008) Free basic water – a sustainable instrument for a sustainable future in South Africa *Environment and Urbanization* 20(1) 67–87. DOI: 10.1177/0956247808089149 Cited on page(s) 68

Naess, A. (1995) The Deep Ecological Movement: Some philosophical aspects, in Sessions G. (ed.) *Deep Ecology for the Twenty First Century* Boston: Shambala, 151–155. Cited on page(s) 37

Naess, A. and Sessions, G. (1995) Platform Principles of the Deep Ecology Movement, in Inoue A. and Inoue Y. (eds) *The Deep Ecology Movement: An Introductory Anthology* Berkeley: North Atlantic Books, 49–53. Cited on page(s) 37

Neumayer, E. (2010) *Weak versus Strong Sustainability* 3rd edition, Cheltenham: Edward Elgar. Cited on page(s) 18, 43

Plumwood, V. (1993) *Feminism and the mastery of nature* New York: Routledge. Cited on page(s) 4, 40

Reaka-Kudla, M., Wilson, D., Wilson, E. (1997) *Biodiversity II* Washington DC: John Henry Press. Cited on page(s) 12

Rifkin, J. (2002) *The Hydrogen Economy* Cambridge: Polity Press. Cited on page(s)

Riley, D. (2008) *Engineering and Social Justice* San Rafael: Morgan and Claypool. DOI: 10.2200/S00117ED1V01Y200805ETS007 Cited on page(s) 35, 43, 77

Rip, A., Misa, T., and Schot, J. (1995) *Managing Technology in Society: The Approach of Constructive Technology Assessment* London and New York: Pinter. Cited on page(s) 81

Salleh, A. (1984) Deeper than Deep Ecology: the ecofeminist connection *Environmental Ethics* 6(4) 335–341. Cited on page(s) 38

Sankoh, O.A. (1996) Making Environmental Impact Assessment Convincible to Developing Countries *Journal of Environmental Management* 47 185–189. DOI: 10.1006/jema.1996.0044 Cited on page(s)

Schumacher, E.F. (1974) *Small is Beautiful* 2nd edition, London: Abacus. Cited on page(s) 18, 58

Shove, E. (2003) *Comfort, Cleanliness and Convenience* Oxford: Berg. Cited on page(s) 53, 54, 56, 57, 64, 67

Sofoulis, Z. (2005) Big Water, Everyday Water: A Sociotechnical Perspective *Continuum: Journal of Media and Cultural Studies* 19(4) 445–463. DOI: 10.1080/10304310500322685 Cited on page(s) 57, 67

United Nations (UN) (2011) *Millennium Development Goals Report* New York: United Nations. Cited on page(s) 9, 20

United Nations Commission for Sustainable Development (UNCSD) (1999) *Comprehensive Assessment of the Freshwater Resources of the World* New York: UN Division for Sustainable Development. Cited on page(s) 8

United Nations Department of Economic and Social Affairs (UNDESA) (2004) *World Population to 2300* New York: United Nations. Cited on page(s) 8

United Nations Development Programme (2006) *Beyond scarcity: Power, poverty and the global water crisis. Human Development Report* New York: Palgrave Macmillan. Cited on page(s) 10

United Nations Environment Programme (UNEP) (2007) *Global Environmental Outlook Geo 4* Valletta, Malta: Progress Press. Cited on page(s) 10, 11

United Nations Human Settlements Programme (UN Habitat) (2006) *The state of the world's cities 2006/07* London: Earthscan. Cited on page(s) 8, 18

United Nations Water (UN Water) (2006) *Coping with Water Scarcity* online `http://www.unwater.org/downloads/waterscarcity.pdf`, accessed 22 July 2011. Cited on page(s) 9

United States Environmental Protection Agency (USEPA) (2002) *Case studies in water conservation* USEPA Office of Water, on-line `http://www.epa.gov/WaterSense/docs/utilityconservation_508.pdf`, accessed 25 July 2011. Cited on page(s) 66

Van Der Ryn, S. and Cowan, S. (2007) *Ecological Design* 10th anniversary edition, Washington: Island Press. Cited on page(s)

van Vliet, B., Chappells, H., and Shove, E. (2005) *Infrastructures of Consumption* London: Earthscan Cited on page(s) 54, 59, 69

Vermaas, P., Kroes, P., van de Poel, I., Franssen, M., and Houkes, W. (2011) *A Philosophy of Technology – From Technical Artefacts to Sociotechnical Systems* San Rafael: Morgan and Claypool. DOI: 10.2200/S00321ED1V01Y201012ETS014 Cited on page(s) 3, 75, 76, 79

Warren, K. (1987) Feminism and ecology: making connections, *Environmental Ethics* 9(1) 3–20. Cited on page(s) 40

Wathern, P. (1988) *Environmental Impact Assessment: Theory and Practice* Sydney: Unwin Hyman Ltd. Cited on page(s)

White, L.T. (1967) The Historical Roots of Our Ecologic Crisis *Science* 155(3767)1203–1207. DOI: 10.1126/science.155.3767.1203 Cited on page(s)

White, T. (2003) Domination, resistance and accommodation in China's one child campaign, in Perry E. and Selden M. *Chinese Society* 2nd edition, London: Routledge Curson. Cited on page(s) 16

Wilson, E. (1992) *The Diversity of Life* Cambridge MA: Belknap. Cited on page(s) 12

World Commission on Environment and Development (WCED) (1987) *Our Common Future* Oxford: Oxford University Press. Cited on page(s) 19

World Commission on Dams (WCD) (2000) *Dams and development: a new framework for decision-making* London: Earthscan. Cited on page(s) 62

Yiatros, S., Waddee, M.A., and Hunt, G.R. (2007) The load-bearing duct: biomimicry in structural design *Proceedings of the Institution of Civil Engineers: Engineering Sustainability* DOI: 10.1680/ensu.2007.160.4.179 Cited on page(s) 2

Author's Biography

SARAH BELL

Sarah Bell is a Chartered Engineer and academic staff member of the Department of Civil, Environmental and Geomatic Engineering at University College London. She began her career as a process and environmental engineer at an aluminium smelter in Australia before completing her PhD in Sustainability and Technology Policy at Murdoch University. She taught sustainable agriculture and land management at the University of Sydney before moving to UCL where she teaches engineering history, environmental systems engineering and an introduction to engineering for liberal arts students. Her research focuses on the interactions between society and technology as they impact on sustainability, with particular focus on urban water systems.

Printed in the United States
by Baker & Taylor Publisher Services